Ring Theory

MONOGRAPHS AND TEXTBOOKS IN PURE AND APPLIED MATHEMATICS

1. K. Yano. Integral Formulas in Riemannian Geometry (1970)
2. S. Kobayashi. Hyperbolic Manifolds and Holomorphic Mappings (1970)
3. V. S. Vladimirov. Equations of Mathematical Physics (A. Jeffrey, editor; A. Littlewood, translator) (1970)
4. B. N. Pshenichnyi. Necessary Conditions for an Extremum (L. Neustadt, translation editor; K. Makowski, translator) (1971)
5. L. Narici, E. Beckenstein, and G. Bachman. Functional Analysis and Valuation Theory (1971)
6. D. S. Passman. Infinite Group Rings (1971)
7. L. Dornhoff. Group Representation Theory (in two parts). Part A: Ordinary Representation Theory. Part B: Modular Representation Theory (1971, 1972)
8. W. Boothby and G. L. Weiss (eds.). Symmetric Spaces: Short Courses Presented at Washington University (1972)
9. Y. Matsushima. Differentiable Manifolds (E. T. Kobayashi, translator) (1972)
10. L. E. Ward, Jr. Topology: An Outline for a First Course (1972)
11. A. Babakhanian. Cohomological Methods in Group Theory (1972)
12. R. Gilmer. Multiplicative Ideal Theory (1972)
13. J. Yeh. Stochastic Processes and the Wiener Integral (1973)
14. J. Barros-Neto. Introduction to the Theory of Distributions (1973)
15. R. Larsen. Functional Analysis: An Introduction (1973)
16. K. Yano and S. Ishihara. Tangent and Cotangent Bundles: Differential Geometry (1973)
17. C. Procesi. Rings with Polynomial Identities (1973)
18. R. Hermann. Geometry, Physics, and Systems (1973)
19. N. R. Wallach. Harmonic Analysis on Homogeneous Spaces (1973)
20. J. Dieudonné. Introduction to the Theory of Formal Groups (1973)
21. I. Vaisman. Cohomology and Differential Forms (1973)
22. B.-Y. Chen. Geometry of Submanifolds (1973)
23. M. Marcus. Finite Dimensional Multilinear Algebra (in two parts) (1973, 1975)
24. R. Larsen. Banach Algebras: An Introduction (1973)
25. R. O. Kujala and A. L. Vitter (eds). Value Distribution Theory: Part A; Part B. Deficit and Bezout Estimates by Wilhelm Stoll (1973)
26. K. B. Stolarsky. Algebraic Numbers and Diophantine Approximation (1974)
27. A. R. Magid. The Separable Galois Theory of Commutative Rings (1974)
28. B. R. McDonald. Finite Rings with Identity (1974)
29. I. Satake. Linear Algebra (S. Koh, T. Akiba, and S. Ihara, translators) (1975)
30. J. S. Golan. Localization of Noncommutative Rings (1975)
31. G. Klambauer. Mathematical Analysis (1975)
32. M. K. Agoston. Algebraic Topology: A First Course (1976)
33. K. R. Goodearl. Ring Theory: Nonsingular Rings and Modules (1976)

Ring Theory

NONSINGULAR RINGS AND MODULES

K. R. Goodearl

Department of Mathematics
University of Utah
Salt Lake City, Utah

MARCEL DEKKER, INC. New York and Basel

MARCEL DEKKER, INC.
270 Madison Avenue, New York, New York 10016

LIBRARY OF CONGRESS CATALOG CARD NUMBER:
75–25169

ISBN: 0–8247–6354–8
Current printing (last digit):
10 9 8 7 6 5 4 3 2 1

PRINTED IN THE UNITED STATES OF AMERICA

Preface

This book is an account of a certain portion of noncommutative ring theory which originated in the work of R. E. Johnson [113, 115], who introduced nonsingular rings, in 1951, and nonsingular modules, in 1957. These concepts proved particularly useful in the general theory of quotient rings initiated by Y. Utumi [195] in 1956. As the use of maximal quotient rings has become fairly widespread in noncommutative ring theory, so has the use of nonsingular rings and modules. However, except for a treatment of some aspects of this theory in the author's memoir [78], nonsingular rings and modules have only been treated in small pieces in widely scattered sources. This prompted the writing of the present book, which aims at a unified exposition of the majority of the ideas and results presently associated with nonsingular rings and modules.

The material presented in this book was chosen with the purpose of making this a useful reference for researchers in noncommutative ring theory, but the presentation itself has been expanded somewhat beyond the needs of a reference work in order to make the book suitable as a text for a graduate course or seminar. For this purpose, over 460 exercises have been included, ranging from trivial consequences of the definitions to difficult problems. (A few of the exercises relate material in the book to ideas not discussed here, but these are easy to spot.) None of the proofs of theorems in the text are relegated to exercises, in order that this book might be a reference for proofs as well as theorems. Also included in the book are a large number of examples designed to illustrate the material. Some of these are presented in the text, some are left as exercises. In addition, Chapter 4 provides a discussion of three general techniques which have proved useful for constructing certain kinds of examples.

Those definitions and results from general ring and module theory needed in this book are listed in the Background Sketch which precedes Chapter 1. The variety in this list suggests that this book might also be studied as an illustration of some of the applications of various standard topics from the theory of rings and modules. Not everything given in the Background Sketch is needed throughout the book. For instance, composition series are used only in Section 5.C, and cardinal numbers are used only in Section 6.C. A small amount of category theory is used many places in the book, but the only essential use of Morita theory occurs in Section 5.C. (Other references to Morita-equivalent rings may be omitted if desired.) As far as homological algebra is concerned, only the basic properties of projective, injective, and flat modules, and hereditary, semihereditary, and regular rings, are needed in the text. The use of Ext, Tor, and homological dimension has been kept out of the text, but does appear in a few exercises.

For use in a course with limited time, some of the topics in the book may be omitted. Possible goals for such a short course might be Goldie's Theorems (Theorem 3.35 and Corollary 3.36), or the nonsingular analogue of the Wedderburn-Artin Theorem (Theorem 5.28). In such a case, the following material may be omitted: all of Section 2.B except Corollary 2.22 (which is not hard to prove directly from the definitions); all of Section 2.D except Lemma 2.33 (which is only needed in Lemma 5.26 and Theorems 6.1 and 6.3); Propositions 3.21 to 3.23 and 3.26 to 3.29; all of Section 4.A except Corollary 4.9 (which is only needed for Proposition 5.22 and thus might be presented without proof); all of Sections 4.B and 4.C.

There are no hidden hypotheses in any of the theorems of this book, except for the usual convention that rings are associative with 1 and that modules are unital. In particular, there are no universal assumptions such as "all rings are nonsingular." On the contrary, hypotheses are as general as possible without undue complication, and nonsingularity is included in hypotheses only when necessary. The phrase "as general as possible" is used here in the context of the theory of rings and modules, rather than in the context of category theory. While a fair proportion of this book could be worked out in a Grothendieck category, as large a proportion is strictly ring theory. For this reason, everything has been stated in the language of rings and modules, but the arguments used are categorical wherever possible. Thus the interested reader can make the necessary translations to category theory if he so desires.

In the interest of compactness, some notation has been abbreviated, and it is hoped that no confusion will result. For example, the word "map" and the symbol "$f: A \to B$" are assumed to refer to module homomorphisms unless otherwise specified. The natural map from a module A to a factor module A/B is given by the rule $x \mapsto \bar{x}$ whenever elements are required in the discussion (i.e., $\bar{x}$ stands for the coset $x + B$). A direct sum of a family of modules $\{A_\alpha | \alpha \in X\}$ is written $\oplus A_\alpha$ whenever the index set X is not pertinent to the discussion.

The material presented in this book is organized into six chapters, each divided into three or four sections. There is a Background Sketch preceding Chapter 1, and some Historical Notes following Chapter 6, along with an extensive bibliography and index. Theorems, propositions, etc., are numbered consecutively throughout each chapter. Thus an Arabic decimal $m.k$ refers to the kth. proposition in Chapter m. The exercises, however, are only numbered consecutively within each section. Thus Exercise n in Section M of Chapter X is referred to as "Exercise n" within that section, and as "Exercise $X.M.n$" elsewhere in the book.

I would like to thank Ann K. Boyle, William D. Emerson, and Robert B. Warfield, Jr., for many pleasant discussions involving the material in this book, and for their comments on portions of the manuscript.

K. R. Goodearl

Contents

Ring Theory

Background Sketch

This section consists of a list of definitions and properties, loosely organized into seven groups, which the reader should consult for unfamiliar terminology, notation, and theorems. Details may be found in standard texts on algebra and ring theory, for example, Anderson-Fuller [5], Faith [52], Hungerford [100], Jacobson [111], Lambek [128], MacLane-Birkhoff [137], Popescu [160].

Rings

All rings in this book are associative with unit, and subrings of a ring R are assumed to have the same unit as R. A *ring morphism* (*ring map*, *ring homomorphism*) from a ring R to a ring S is a function $R \to S$ which preserves addition, multiplication, and 1. Unless otherwise specified, the letter R always stands for a ring. If I is a two-sided ideal of R, we sometimes use the notation $x \mapsto \bar{x}$ for the natural ring morphism $R \to R/I$. The *opposite ring* of R is the ring R^{op} consisting of the same set of elements as R and the same addition, but with multiplication defined by the rule $r * s = sr$.

An element $x \in R$ is *central* if $xr = rx$ for all $r \in R$. The *center* of R is the set of all central elements of R and is a subring of R.

A *simple ring* is a nonzero ring R in which the only two-sided ideals are 0 and R.

The *right* (*left*) *annihilator* of a subset X of R is the set of those $r \in R$ such that $xr = 0$ ($rx = 0$) for all $x \in X$.

An *idempotent* in R is an element $e \in R$ such that $e^2 = e$. Two idempotents

$e,f \in R$ are *orthogonal* if $ef = fe = 0$. A set $\{e_\alpha\}$ of idempotents in R is said to be (*pairwise*) *orthogonal* if e_α is orthogonal to e_β whenever $\alpha \neq \beta$.

Proposition If e is an idempotent in R, then eRe is a ring with unit e. Also, eRe is isomorphic to the endomorphism ring of the right ideal eR (provided endomorphisms of eR are written on the left of their arguments).□

Proposition (a) If $e_1, \ldots, e_n$ are orthogonal idempotents in R such that $e_1 + \cdots + e_n = 1$, then $R = e_1R \oplus \cdots \oplus e_nR$.

(b) If $R = I_1 \oplus \cdots \oplus I_n$ for some right ideals $I_1, \ldots, I_n$, then there exist orthogonal idempotents $e_1, \ldots, e_n \in R$ such that $e_1 + \cdots + e_n = 1$ and $e_jR = I_j$ for all j.

(c) If $R = I_1 \oplus \cdots \oplus I_n$ for some two-sided ideals $I_1, \ldots, I_n$, then there exist orthogonal central idempotents $e_1, \ldots, e_n \in R$ such that $e_1 + \cdots + e_n = 1$ and $e_jR = I_j$ for all j.□

A ring R is *indecomposable* (as a ring) if the only ring decompositions $R \cong R_1 \times R_2$ are those in which either $R_1 = 0$ or $R_2 = 0$. Equivalently, R is indecomposable if and only if the only central idempotents in R are 0 and 1.

An element $x \in R$ (or an ideal I of R) is *nilpotent* if there exists a positive integer n such that $x^n = 0$ ($I^n = 0$). A subset X of R is *nil* if every element of X is nilpotent.

A *prime ideal* in R is a proper two-sided ideal P such that whenever $a,b \in R$ with $aRb \subseteq P$, either $a \in P$ or $b \in P$. A *prime ring* is a ring in which 0 is a prime ideal. Thus P is a prime ideal if and only if R/P is a prime ring.

Proposition If P is a proper two-sided ideal of R, then the following conditions are equivalent:

(a) P is a prime ideal.

(b) If I,J are right ideals of R such that $IJ \subseteq P$, then either $I \subseteq P$ or $J \subseteq P$.

(c) If I,J are left ideals of R such that $IJ \subseteq P$, then either $I \subseteq P$ or $J \subseteq P$.

(d) If I,J are two-sided ideals of R which properly contain P, then $IJ \nsubseteq P$.□

A *semiprime ideal* in R is a two-sided ideal I such that whenever $a \in R$ with $aRa \subseteq I$, then $a \in I$. A *semiprime ring* is a ring in which 0 is a semiprime ideal. Thus I is a semiprime ideal if and only if R/I is a semiprime ring.

Proposition If I is a two-sided ideal of R, then the following conditions are equivalent:

(a) I is a semiprime ideal.

(b) R/I has no nonzero nilpotent right ideals.
(c) R/I has no nonzero nilpotent left ideals.
(d) If J is a two-sided ideal of R which properly contains I, then $J^2 \nsubseteq I$.
(e) I is an intersection of prime ideals.□

Proposition (a) Every prime ideal (ring) is also a semiprime ideal (ring).
(b) Every simple ring is a prime ring.
(c) Every maximal two-sided ideal is a prime ideal.□

The *prime radical* of R is the intersection of all prime ideals of R. Thus R is a semiprime ring if and only if its prime radical is 0.

Proposition Let P denote the prime radical of R.
(a) P is a semiprime ideal which is contained in every semiprime ideal of R.
(b) P is a nil two-sided ideal which contains every nilpotent right and left ideal of R.
(c) The prime radical of R/P is 0.□

A *right* (*left*) *primitive ideal* in R is a two-sided ideal P such that there exists a maximal right (left) ideal M with $P = \{x \in R \mid Rx \subseteq M\}$ ($P = \{x \in R \mid xR \subseteq M\}$), i.e., such that P is the largest two-sided ideal of R which is contained in M. A *right* (*left*) *primitive ring* is a ring in which 0 is a right (left) primitive ideal. Thus P is a right (left) primitive ideal if and only if R/P is a right (left) primitive ring.

Proposition (a) Every right or left primitive ideal (ring) is also a prime ideal (ring).
(b) Every simple ring is a right and left primitive ring.
(c) Every maximal two-sided ideal is a right and left primitive ideal.□

An element $x \in R$ is *right* (*left*) *quasi-regular* if $1 - x$ has a right (left) inverse in R. Also, x is *quasi-regular* if $1 - x$ is invertible in R. A subset X of R is *right* (*left*) *quasi-regular* if every element of X is right (left) quasi-regular. Also, X is *quasi-regular* if every element of X is quasi-regular. The *Jacobson radical* of R is the set

$$J(R) = \{x \in R \mid \text{xR is right quasi-regular}\}$$

Theorem (a) $J(R) = \{x \in R \mid Rx \text{ is left quasi-regular}\}$.
(b) $J(R)$ is a quasi-regular two-sided ideal of R which contains every right (left) quasi-regular right (left) ideal of R.
(c) $J(R)$ equals the intersection of all maximal right (left) ideals of R.

(d) $J(R)$ equals the intersection of all right (left) primitive ideals of R.□

Proposition (a) $J(R/J(R)) = 0$.
(b) $J(R)$ contains every nil right and left ideal of R.
(c) $J(R)$ contains the prime radical of R.□

Modules

All modules in this book are *unital*: that is, if A is a right (left) R-module, then $a1 = a$ ($1a = a$) for all $a \in A$. If A is a module, then the notation $B \leqq A$ means that B is a submodule of A and the notation $B < A$ means that B is a proper submodule of A. In particular, the notation $I \leqq R_R$ ($I \leqq {}_RR$) means that I is a right (left) ideal of R. When $B \leqq A$, we sometimes use the notation $x \mapsto \bar{x}$ for the natural homomorphism $A \to A/B$.

Given rings R and S, an *R,S-bimodule* is an abelian group A which is both a left R-module and a right S-module in such a way that $(ra)s = r(as)$ for all $r \in R$, $a \in A$, $s \in S$.

A right (left) R-module A is *faithful* if for any nonzero $r \in R$ there is an element $a \in A$ such that $ar \neq 0$ ($ra \neq 0$). Equivalently, A is faithful if and only if R_R (${}_RR$) can be embedded in some direct product of copies of A.

A module A is *indecomposable* if the only direct sum decompositions $A = A_1 \oplus A_2$ are those in which either $A_1 = 0$ or $A_2 = 0$.

Most modules in this book are right modules, and so homomorphisms are written on the *left* of their arguments. If A and B are modules, then the phrase "map from A to B," or the notation "$f : A \to B$," refers to a homomorphism unless otherwise specified.

The ring of all endomorphisms of an R-module A is denoted $\mathrm{End}_R(A)$. A submodule $B \leqq A$ is *fully invariant* in A if $fB \leqq B$ for all $f \in \mathrm{End}_R(A)$.

If X is any family of submodules of a module A, we use ΣX to stand for the set of all sums $a_1 + \cdots + a_n$ such that there exist $A_1, \ldots, A_n \in X$ with $a_i \in A_i$ for all i. Equivalently, ΣX is the submodule of A generated by the submodules in X. We say that X is *independent* if $A_0 \cap (A_1 + \cdots + A_n) = 0$ for all distinct $A_0, A_1, \ldots, A_n \in X$. Equivalently, X is independent if and only if $(\Sigma X_1) \cap (\Sigma X_2) = 0$ for all disjoint subsets $X_1, X_2 \subseteq X$. Also, X is independent if and only if the natural map $\oplus X \to \Sigma X$ is an isomorphism, in which case we identify $\oplus X$ with ΣX.

Proposition (a) Let $\{X_\alpha\}$ be a collection of independent families of submodules of A. If $\{\Sigma X_\alpha\}$ is an independent family of submodules of A, then $\cup X_\alpha$ is independent.

(b) A sequence $\{A_1, A_2, \ldots\}$ of submodules of A is independent if and only if $A_{n+1} \cap (A_1 + \cdots + A_n) = 0$ for all n.□

A *simple module* (*irreducible module*) is a nonzero module A in which the only submodules are 0 and A. Equivalently, a right (left) R-module A is simple if and only if $A \cong R/M$ for some maximal right (left) ideal M of R.

Schur's Lemma If A is a simple R-module, then $\mathrm{End}_R(A)$ is a division ring.□

Domains

An element $x \in R$ is a *zero-divisor* if there exists a nonzero element $r \in R$ such that either $rx = 0$ or $xr = 0$. An *integral domain* is a nonzero ring R (not necessarily commutative) in which the only zero-divisor is 0. Equivalently, a nonzero ring R is an integral domain if and only if $ab \neq 0$ for all nonzero $a,b \in R$.

If R is a commutative integral domain, then the *quotient field* of R is a field K with R as a subring such that every element of K can be expressed in the form rs^{-1} for suitable $r,s \in R$.

Let R be a commutative integral domain, and let A be an R-module. The set

$$T(A) = \{a \in A \mid ar = 0 \text{ for some nonzero } r \in R\}$$

is a submodule of A, called the *torsion submodule* of A. Then A is a *torsion module* if $T(A) = A$, while A is a *torsion-free module* if $T(A) = 0$.

We use $\mathbf{Z}$ to denote the commutative integral domain consisting of the integers, and $\mathbf{Q}$ to denote the field of rational numbers, which is the quotient field of $\mathbf{Z}$. Given any prime number p, we use $\mathbf{Z}(p^\infty)$ to denote the $\mathbf{Z}$-module

$$\frac{\{m/p^n \mid m,n \in \mathbf{Z},\, n > 0\}}{\mathbf{Z}} \leqq \frac{\mathbf{Q}}{\mathbf{Z}}$$

Order

A *partial order* on a set X is a relation $\leqq$ such that (a) $x \leqq x$ for all $x \in X$; (b) if $x,y \in X$ with $x \leqq y$ and $y \leqq x$, then $x = y$; (c) if $x,y,z \in X$ with $x \leqq y$ and $y \leqq z$, then $x \leqq z$.

A *poset* (*partially ordered set*) is a set P together with a specified partial order $\leqq$ on P. In a poset P, we write $x \geqq y$ to mean that $y \leqq x$, and $x < y$ (or $y > x$) to mean that $x \leqq y$ but $x \neq y$.

Let X be a subset of a poset P. An *upper bound* (*lower bound*) for X is any element $a \in P$ such that $a \geqq x$ ($a \leqq x$) for all $x \in X$. A *least upper bound*, or *supremum* (*greatest lower bound*, or *infimum*) of X is an upper bound (lower bound) a for X such that $a \leqq b$ ($a \geqq b$) for all upper bounds (lower bounds) b of X. If X has a supremum (infimum), it is unique and is denoted $\vee X$ ($\wedge X$). The supremum (infimum) of a two-element set $\{x,y\}$ is denoted $x \vee y$ ($x \wedge y$), if it exists.

A *lattice* is a poset L such that $x \vee y$ and $x \wedge y$ exist for all $x,y \in L$. A *complete lattice* is a poset L such that $\vee X$ and $\wedge X$ exist for all subsets X of L. If L is a complete lattice, then L has a least element and a greatest element (namely $\wedge L$ and $\vee L$), which are usually denoted 0 and 1. A lattice L is *modular* if it satisfies the *modular law*: Whenever $a,b,c \in L$ with $a \leqq c$, then $a \vee (b \wedge c) = (a \vee b) \wedge c$.

Proposition The collection of all submodules of a module A, using the inclusion relation $\leqq$, is a complete modular lattice. □

If P and Q are posets, a function $f : P \to Q$ is *monotone* (*order-preserving*) if whenever $x \leqq y$ in P, then $fx \leqq fy$ in Q. An *order isomorphism* of P onto Q is a bijection $f : P \to Q$ such that f and f^{-1} are both monotone. Equivalently, a bijection $f : P \to Q$ is an order isomorphism if and only if for all $x,y \in P$, $x \leqq y$ if and only if $fx \leqq fy$.

If L and M are lattices, a *lattice homomorphism* from L to M is a function $f : L \to M$ such that $f(x \vee y) = fx \vee fy$ and $f(x \wedge y) = fx \wedge fy$ for all $x,y \in L$. Every lattice homomorphism is monotone. A *lattice isomorphism* of L onto M is a bijection $f : L \to M$ such that f and f^{-1} are both lattice homomorphisms. Every lattice isomorphism is also an order isomorphism, and every order isomorphism (between lattices) is also a lattice isomorphism.

Two elements x,y in a poset P are *comparable* provided either $x \leqq y$ or $y \leqq x$. A subset X of P is *linearly ordered* (*totally ordered*) if every pair of elements in X are comparable. A linearly ordered subset of P is also called a *chain*. Every chain is also a lattice.

A *maximal element* (*minimal element*) in a subset X of a poset P is an element $x \in X$ such that there are no elements $y \in X$ with $y > x$ ($y < x$). A *greatest element* (*least element*) in X is an element $x \in X$ such that $x \geqq y$ ($x \leqq y$) for all $y \in X$. We say that P is *well-ordered* if every nonempty subset of P has a least element. Any well-ordered poset which has a greatest element is also a complete lattice.

Zorn's Lemma If P is a nonempty poset such that every nonempty chain in P has an upper bound in P, then P has a maximal element. □

Every set X has a *cardinality*, denoted card(X), such that there is a bijection between two sets X and Y if and only if $\text{card}(X) = \text{card}(Y)$. If there exists an injection $X \to Y$, then we write $\text{card}(X) \leqq \text{card}(Y)$. This defines a partial order on the class of cardinal numbers.

Theorem The class of cardinal numbers is well-ordered.□

The *successor* of a cardinal α is the smallest cardinal β such that $\beta > \alpha$. A *successor cardinal* is any cardinal which is the successor of some cardinal. Every other cardinal (except 0) is called a *limit cardinal*. Equivalently, a cardinal α is a limit cardinal if and only if there exists a nonempty set X of cardinals such that $\beta < \alpha$ for all $\beta \in X$ and $\alpha = \vee X$.

Given cardinals α and β, we choose disjoint sets X and Y such that $\text{card}(X) = \alpha$ and $\text{card}(Y) = \beta$, and we define $\alpha + \beta = \text{card}(X \cup Y)$ and $\alpha\beta = \text{card}(X \times Y)$.

Theorem Let α and β be nonzero cardinals. If either α or β is infinite, then $\alpha + \beta = \alpha\beta = \max\{\alpha,\beta\}$.□

Chain Conditions

A poset P satisfies the *ascending chain condition*, or *ACC* (*descending chain condition*, or *DCC*) if every chain in P of the form $x_1 \leqq x_2 \leqq \cdots$ $(x_1 \geqq x_2 \geqq \cdots)$ is ultimately constant, i.e., if there is a positive integer N such that $x_n = x_N$ for all $n \geqq N$. Equivalently, P satisfies the ACC (DCC) if and only if every nonempty subset of P has a maximal (minimal) element.

A module A is *noetherian* (*artinian*) if the lattice of submodules of A satisfies the ACC (DCC).

Proposition Given a module A, the following conditions are equivalent:

(a) A is noetherian.
(b) All submodules of A are finitely generated.
(c) The finitely generated submodules of A satisfy the ACC.□

Proposition Let $B \leqq A$ be modules. Then A is noetherian (artinian) if and only if B and A/B are both noetherian (artinian).□

The ring R is *right* (*left*) *noetherian* (*artinian*) if R_R $({}_RR)$ is noetherian (artinian).

Proposition If R is right noetherian (artinian), then all finitely generated right R-modules are noetherian (artinian).□

Hilbert Basis Theorem If R is a right noetherian ring, then $R[x]$ is right noetherian.□

A *composition series* for a module A is a chain of submodules $A_0 = 0 < A_1 < \cdots < A_n = A$ such that A_i/A_{i-1} is simple for $i = 1, \ldots, n$. The integer n is called the *length* of the composition series, and the modules A_i/A_{i-1} are called the *composition factors* of the composition series.

Proposition A module has a composition series if and only if it is both noetherian and artinian.□

Jordan-Hölder Theorem If a module has two composition series, then the composition series have the same length and there is a bijection between the two sets of composition factors such that corresponding composition factors are isomorphic.□

Homology

A sequence $\cdots \to A_{n-1} \xrightarrow{f_{n-1}} A_n \xrightarrow{f_n} A_{n+1} \to \cdots$ of modules and homomorphisms is *exact* if $\ker f_n = f_{n-1}A_{n-1}$ for all n. A *short exact sequence* is any exact sequence of the form $0 \to A \to B \to C \to 0$. A short exact sequence $0 \to A \xrightarrow{f} B \xrightarrow{g} C \to 0$ is said to *split* if there exists a map $h: B \to A$ such that $hf = 1_A$, or, equivalently, if there exists a map $k : C \to B$ such that $gk = 1_C$.

Proposition Let $0 \to A \to B \to C \to 0$ be any short exact sequence of right R-modules.

(a) For any right R-module M, the induced sequences

$$0 \to \mathrm{Hom}_R(M,A) \to \mathrm{Hom}_R(M,B) \to \mathrm{Hom}_R(M,C)$$

and

$$0 \to \mathrm{Hom}_R(C,M) \to \mathrm{Hom}_R(B,M) \to \mathrm{Hom}_R(A,M)$$

are exact sequences of **Z**-modules.

(b) For any left R-module N, the induced sequence

$$A \otimes_R N \to B \otimes_R N \to C \otimes_R N \to 0$$

is an exact sequence of **Z**-modules.□

A module A is *finitely presented* (*finitely related*) if there exists a short exact sequence $0 \to K \to F \to A \to 0$ such that F is finitely generated free and K is finitely generated.

Proposition Let A be a finitely presented module. If $0 \to C \to B \to A \to 0$ is any short exact sequence with B finitely generated, then C is finitely generated.□

A module P is *projective* if given any epimorphism $f : A \to B$ and any homomorphism $g : P \to B$, there exists a homomorphism $h : P \to A$ such that $fh = g$.

Proposition Given a module P, the following conditions are equivalent:

(a) P is projective.

(b) For every short exact sequence $0 \to A \to B \to C \to 0$, the induced sequence $0 \to \mathrm{Hom}_R(P,A) \to \mathrm{Hom}_R(P,B) \to \mathrm{Hom}_R(P,C) \to 0$ is exact.

(c) Every short exact sequence $0 \to A \to B \to P \to 0$ splits.

(d) P is isomorphic to a direct summand of a free module.□

Dual Basis Lemma A right R-module P is projective if and only if there exist elements $\{x_\lambda \mid \lambda \in L\}$ in P and homomorphisms $\{f_\lambda \mid \lambda \in L\}$ in $\mathrm{Hom}_R(P,R_R)$ such that for each $x \in P$, $f_\lambda x = 0$ for all but finitely many $\lambda \in L$, and $x = \Sigma x_\lambda(f_\lambda x)$.□

Proposition (a) A module $P = \oplus P_\alpha$ is projective if and only if each P_α is projective.

(b) Every module is an epimorphic image of a projective module.□

Schanuel's Lemma Let $0 \to K \to P \to A \to 0$ and $0 \to L \to Q \to A \to 0$ be exact. If P and Q are projective, then $K \oplus Q \cong L \oplus P$.□

A module I is *injective* if given any monomorphism $f : A \to B$ and any homomorphism $g : A \to I$, there exists a homomorphism $h : B \to I$ such that $hf = g$.

Proposition Given a module I, the following conditions are equivalent:

(a) I is injective.

(b) For every short exact sequence $0 \to A \to B \to C \to 0$, the induced sequence $0 \to \mathrm{Hom}_R(C,I) \to \mathrm{Hom}_R(B,I) \to \mathrm{Hom}_R(A,I) \to 0$ is exact.

(c) Every short exact sequence $0 \to I \to B \to C \to 0$ splits.□

Baer's Criterion A right R-module I is injective if and only if for each right ideal J of R, every homomorphism $J \to I$ extends to a homomorphism $R_R \to I$.□

Proposition A module $I = \Pi I_\alpha$ is injective if and only if each I_α is injective.□

Theorem Every module is a submodule of an injective module. □

Theorem A ring R is right noetherian if and only if all direct sums of injective right R-modules are injective. □

The ring R is *right* (*left*) *self-injective* provided R_R (${}_RR$) is injective.

Proposition A ring $R = \Pi R_\alpha$ is right (left) self-injective if and only if each R_α is right (left) self-injective. □

A right (left) R-module F is *flat* if given any monomorphism $A \to B$ of left (right) R-modules, the induced homomorphism $F \otimes_R A \to F \otimes_R B$ ($A \otimes_R F \to B \otimes_R F$) is also a monomorphism.

Proposition (a) A right (left) R-module F is flat if and only if, for every left (right) ideal J of R, the natural map $F \otimes_R J \to FJ$ ($J \otimes_R F \to JF$) is an isomorphism of $\mathbf{Z}$-modules.

(b) A module $F = \oplus F_\alpha$ is flat if and only if each F_α is flat.

(c) Every projective module is flat.

(d) If all finitely generated submodules of a module F are flat, then F is flat. □

The ring R is (*von Neumann*) *regular* if for every $a \in R$ there is an $x \in R$ such that $axa = a$.

Theorem The following conditions are equivalent:

(a) R is regular.

(b) Every principal right (left) ideal of R is generated by an idempotent.

(c) Every finitely generated submodule of a projective right (left) R-module is a direct summand.

(d) Every right (left) R-module is flat. □

The ring R is *right* (*left*) *hereditary* if every right (left) ideal of R is projective. Similarly, R is *right* (*left*) *semihereditary* if every finitely generated right (left) ideal of R is projective.

Theorem Let R be right hereditary.

(a) Every submodule of a projective right R-module is projective.

(b) Every factor module of an injective right R-module is injective.

(c) Every projective right R-module is isomorphic to a direct sum of copies of right ideals of R. □

Theorem Let R be right semihereditary.

(a) Every finitely generated submodule of a projective right R-module is projective.

(b) Every submodule of a flat right or left R-module is flat.

(c) Every finitely generated projective right R-module is isomorphic to a direct sum of copies of right ideals of R.□

Categories

A *category* $\mathcal{M}$ consists of

(a) A class obj($\mathcal{M}$), whose members are called the *objects* of $\mathcal{M}$.
(b) For each ordered pair (A,B) of objects of $\mathcal{M}$, a set $\mathrm{Hom}_{\mathcal{M}}(A,B)$, whose elements are called *morphisms from A to B.*
(c) For each ordered triple (A,B,C) of objects of $\mathcal{M}$, a composition rule

$$\mathrm{Hom}_{\mathcal{M}}(B,C) \times \mathrm{Hom}_{\mathcal{M}}(A,B) \rightarrow \mathrm{Hom}_{\mathcal{M}}(A,C)$$

where the product of $g \in \mathrm{Hom}_{\mathcal{M}}(B,C)$ with $f \in \mathrm{Hom}_{\mathcal{M}}(A,B)$ is denoted gf.
(d) An *associativity axiom*: $(hg)f = h(gf)$ whenever $f \in \mathrm{Hom}_{\mathcal{M}}(A,B)$, $g \in \mathrm{Hom}_{\mathcal{M}}(B,C)$, and $h \in \mathrm{Hom}_{\mathcal{M}}(C,D)$.
(e) An *identity axiom*: For each object A of $\mathcal{M}$, there must be an *identity morphism* $1_A \in \mathrm{Hom}_{\mathcal{M}}(A,A)$ such that $f1_A = f$ for all $f \in \mathrm{Hom}_{\mathcal{M}}(A,B)$ and $1_A g = g$ for all $g \in \mathrm{Hom}_{\mathcal{M}}(C,A)$.

Given any ring R, there is a category Mod-R (R-Mod) whose objects are all right (left) R-modules, whose morphisms are all right (left) R-module homomorphisms, and whose composition rule is just composition of functions.

A morphism $f \in \mathrm{Hom}_{\mathcal{M}}(A,B)$ in a category $\mathcal{M}$ is an *isomorphism* if there exists a morphism $g \in \mathrm{Hom}_{\mathcal{M}}(B,A)$ such that $fg = 1_B$ and $gf = 1_A$.

A *generator* in a category $\mathcal{M}$ is an object G such that for any distinct morphisms $f,g \in \mathrm{Hom}_{\mathcal{M}}(A,B)$, there exists a morphism $h \in \mathrm{Hom}_{\mathcal{M}}(G,A)$ with $fh \neq gh$.

Proposition Given a right R-module G, the following conditions are equivalent:

(a) G is a generator in Mod-R.

(b) Every right R-module is an epimorphic image of a direct sum of copies of G.

(c) There is a positive integer n such that R_R is isomorphic to a direct summand of G^n.□

A *commutative diagram* in a category $\mathcal{M}$ is a diagram of objects and morphisms (drawn so that a morphism from A to B is represented by an arrow $A \to B$) such that for any objects A and B in the diagram the composition of the morphisms along any path from A to B always yields the same element of $\mathrm{Hom}_{\mathcal{M}}(A,B)$.

A (*covariant*) *functor* from a category $\mathcal{M}$ to a category $\mathcal{K}$ is a function F such that

(a) F maps $\mathrm{obj}(\mathcal{M})$ to $\mathrm{obj}(\mathcal{K})$.
(b) For all $A,B \in \mathrm{obj}(\mathcal{M})$, F maps $\mathrm{Hom}_{\mathcal{M}}(A,B)$ to $\mathrm{Hom}_{\mathcal{K}}(FA,FB)$.
(c) For all $A \in \mathrm{obj}(\mathcal{M})$, $F(1_A) = 1_{FA}$.
(d) For all $f \in \mathrm{Hom}_{\mathcal{M}}(A,B)$ and $g \in \mathrm{Hom}_{\mathcal{M}}(B,C)$, $F(gf) = (Fg)(Ff)$.

A functor $F : \text{Mod-}R \to \text{Mod-}S$ is *additive* if for all $f,g \in \mathrm{Hom}_R(A,B)$, $F(f + g) = Ff + Fg$. Also, F is *exact* if for every short exact sequence $0 \to A \xrightarrow{f} B \xrightarrow{g} C \to 0$ in Mod-R, $0 \to FA \xrightarrow{Ff} FB \xrightarrow{Fg} FC \to 0$ is a short exact sequence in Mod-S. An exact functor also preserves long exact sequences.

Given functors $F,G : \mathcal{M} \to \mathcal{K}$, a *natural transformation* from F to G is a function h which assigns to each $A \in \mathrm{obj}(\mathcal{M})$ a morphism $h(A) \in \mathrm{Hom}_{\mathcal{K}}(FA,GA)$ such that for any morphism $f: A \to B$ in $\mathcal{M}$, $(Gf)(h(A)) = (h(B))(Ff)$. A *natural equivalence* from F to G is a natural transformation $h : F \to G$ such that $h(A)$ is an isomorphism for all $A \in \mathrm{obj}(\mathcal{M})$. If there exists a natural equivalence from F to G, then we say that F and G are *naturally equivalent*.

Two categories $\mathcal{M}$ and $\mathcal{K}$ are *equivalent* if there exist functors $F : \mathcal{M} \to \mathcal{K}$ and $G : \mathcal{K} \to \mathcal{M}$ such that FG is naturally equivalent to the identity functor on $\mathcal{K}$ and GF is naturally equivalent to the identity functor on $\mathcal{M}$. In this case, the functors F and G are called *category equivalences*.

Theorem Any category equivalence Mod-$R \to$ Mod-S preserves:
(a) Zero modules and zero maps
(b) Monomorphisms, epimorphisms, and isomorphisms
(c) Direct sums and products
(d) Exact and split exact sequences
(e) Projective and injective modules
(f) Finitely generated modules
(g) Sums of maps
(h) Endomorphism rings□

Morita's Theorem For any rings R and S, the following conditions are equivalent:
(a) Mod-R is equivalent to Mod-S.

(b) R-Mod is equivalent to S-Mod.

(c) There exists a finitely generated projective generator $P \in$ Mod-R such that $S \cong \mathrm{End}_R(P)$.

(d) There exists a positive integer n and an idempotent e in the ring T of all $n \times n$ matrices over R such that $TeT = T$ and $eTe \cong S$.□

Two rings R and S are *Morita-equivalent* if Mod-R is equivalent to Mod-S. In particular, for any positive integer n, R is Morita-equivalent to the ring of all $n \times n$ matrices over R. A property P of rings is *Morita-invariant* if every ring Morita-equivalent to a ring satisfying P must also satisfy P.

Theorem The following properties are Morita-invariant:

(a) Right (left) hereditary
(b) Right (left) semihereditary
(c) Right (left) noetherian (artinian)
(d) Simple, prime, primitive, semiprime□

1

Essential Extensions and Singular Submodules

This chapter develops several basic concepts which are used throughout the book. The fundamental idea is that of an *essential extension*: A pair of modules $A \leqq B$ such that A intersects every nonzero submodule of B nontrivially. Section A covers the basic properties and existence of essential extensions. The two extremes of essential extensions are covered in Sections B and C. First, every module A has a maximal essential extension, which is the *injective hull* of A. Second, the intersection of all essential submodules of A is the *socle* of A, which involves a discussion of semisimple modules.

Any module isomorphic to the factor B/A of an essential extension $A \leqq B$ is called a *singular module*. Every module A has a unique maximum singular submodule, denoted $Z(A)$, and A is called a *nonsingular module* if $Z(A) = 0$. Section D covers the basic properties of singular and nonsingular modules.

A. Essential Extensions

Definition Consider a submodule A of a module B. We say that B is an *essential extension* of A if every nonzero submodule of B has nonzero intersection with A. We also say that A is an *essential submodule* (or a *large submodule*) of B, and we write $A \leqq_e B$ to denote this situation. In order to test for this condition, we need only check whether all nonzero cyclic submodules of B have nonzero intersection with A, which is equivalent to the condition that every nonzero element of B has a nonzero multiple in A. Note that A always has at least one essential extension, since $A \leqq_e A$. Also, note that $0 \leqq_e A$ only if $A = 0$.

More generally, we use the phrase "essential extension of A" to refer to any monomorphism $f : A \to B$ such that $fA \leqq_e B$. In this case, the map f is called an *essential monomorphism*.

For example, consider the **Z**-submodules of **Q**. Since any two nonzero elements of **Q** have a nonzero common integer multiple, we see that any two nonzero submodules of **Q** have nonzero intersection. Thus if $A \leqq B \leqq \mathbf{Q}$ with $A \neq 0$, we must have $A \leqq_e B$. In particular, **Z** is an essential extension of any of its nonzero ideals, and **Q** is an essential extension of **Z**.

Proposition 1.1 (a) If $A \leqq B \leqq C$, then $A \leqq_e C$ if and only if $A \leqq_e B \leqq_e C$.

(b) If $A \leqq_e B \leqq C$ and $A' \leqq_e B' \leqq C$, then $A \cap A' \leqq_e B \cap B'$.

(c) If $f : B \to C$ and $A \leqq_e C$, then $f^{-1}A \leqq_e B$.

(d) If $\{A_\alpha\}$ is an independent family of submodules of C, and if $A_\alpha \leqq_e B_\alpha \leqq C$ for each α, then $\{B_\alpha\}$ is an independent family and $\oplus A_\alpha \leqq_e \oplus B_\alpha$.

Proof: (a) First let $A \leqq_e B \leqq_e C$ and consider any nonzero $M \leqq C$. Since $B \leqq_e C$ we have $M \cap B \neq 0$, and then since $A \leqq_e B$ we obtain $(M \cap B) \cap A \neq 0$, that is, $M \cap A \neq 0$. Thus $A \leqq_e C$.

Now assume that $A \leqq_e C$. Since any nonzero submodule of C has nonzero intersection with A, the same can be said for nonzero submodules of B; hence $A \leqq_e B$. Also, since any nonzero submodule M of C satisfies $M \cap A \neq 0$, it must satisfy $M \cap B \neq 0$; thus $B \leqq_e C$.

(b) If M is any nonzero submodule of $B \cap B'$, then since $A \leqq_e B$ we have $M \cap A \neq 0$. Since $A' \leqq_e B'$ as well, we obtain $(M \cap A) \cap A' \neq 0$, and thus $A \cap A' \leqq_e B \cap B'$.

(c) If not, then B has a nonzero submodule M such that $M \cap f^{-1}A = 0$. In particular, $M \cap (\ker f) = 0$; hence f maps M isomorphically onto fM, so that fM is a nonzero submodule of C. However, since $M \cap f^{-1}A = 0$ we obtain $fM \cap A = 0$, which is impossible.

(d) First consider the case when the index set consists of exactly two elements, say $\{1,2\}$. According to (b), $0 \leqq_e B_1 \cap B_2$; hence $B_1 \cap B_2 = 0$ and so $\{B_1,B_2\}$ is independent. Applying (c) to the projection maps $B_1 \oplus B_2 \to B_1$ and $B_1 \oplus B_2 \to B_2$, we obtain $A_1 \oplus B_2 \leqq_e B_1 \oplus B_2$ and $B_1 \oplus A_2 \leqq_e B_1 \oplus B_2$; hence it follows from (b) that $A_1 \oplus A_2 \leqq_e B_1 \oplus B_2$.

Thus (d) holds for index sets with two elements. Now consider the case when the index set consists of $\{1,2, \ldots, n\}$, and assume that (d) holds for index sets with $n - 1$ elements. Then $\{B_1, \ldots, B_{n-1}\}$ is independent, and $A_1 \oplus \cdots \oplus A_{n-1} \leqq_e B_1 \oplus \cdots \oplus B_{n-1}$. Using the case above, we see that $(B_1 \oplus \cdots \oplus B_{n-1}) \cap B_n = 0$, whence $\{B_1, \ldots, B_n\}$ is independent, and that

$$(A_1 \oplus \cdots \oplus A_{n-1}) \oplus A_n \leqq_e (B_1 \oplus \cdots \oplus B_{n-1}) \oplus B_n$$

Therefore (d) holds for all finite index sets, and we are ready to prove the general case. Given distinct indices $\alpha(0), \alpha(1), \ldots, \alpha(n)$, we know that

$\{B_{\alpha(0)}, \ldots, B_{\alpha(n)}\}$ is independent, whence $B_{\alpha(0)} \cap (B_{\alpha(1)} + \cdots + B_{\alpha(n)}) = 0$. Thus $\{B_\alpha\}$ is independent. Now any nonzero submodule $M \leqq \oplus B_\alpha$ contains a nonzero element, which must belong to $B_{\alpha(1)} \oplus \cdots \oplus B_{\alpha(n)}$ for some $\alpha(i)$. As a result, $M \cap (B_{\alpha(1)} \oplus \cdots \oplus B_{\alpha(n)}) \neq 0$, from which we obtain

$$M \cap (B_{\alpha(1)} \oplus \cdots \oplus B_{\alpha(n)}) \cap (A_{\alpha(1)} \oplus \cdots \oplus A_{\alpha(n)}) \neq 0$$

and consequently $M \cap (\oplus A_\alpha) \neq 0$. Therefore $\oplus A_\alpha \leqq_e \oplus B_\alpha$.□

We note that 1.1(b) may fail for infinite intersections. For example, $n\mathbf{Z} \leqq_e \mathbf{Z}$ for all positive integers n, and yet $\cap(n\mathbf{Z}) = 0$, which is not essential in $\mathbf{Z}$. Also, 1.1(d) may fail if the family $\{A_\alpha\}$ is not independent, as the following example shows.

Example 1.2 There exist modules $A \leqq B \leqq C$ and $A' \leqq B' \leqq C$ such that $A \leqq_e B$ and $A' \leqq_e B'$, but $A + A' \nleqq_e B + B'$.

Proof: Set $R = \mathbf{Z}$, $C = \mathbf{Z} \oplus (\mathbf{Z}/2\mathbf{Z})$, $A = A' = (2,0)R$, $B = (1,0)R$, and $B' = (1,\bar{1})R$. Any nonzero element of B' has the form $(n,\bar{n})$ for some nonzero $n \in \mathbf{Z}$, and $(n,\bar{n})2 = (2n,0)$ is a nonzero element of A'. Thus $A' \leqq_e B'$, and similarly $A \leqq_e B$. Observing that $(0,\bar{1})R \cap A = 0$, we see that $A \nleqq_e C$, that is, $A + A' \nleqq_e B + B'$.□

Definition Let A be a submodule of C. A *relative complement for A in C* is anysub module B of C which is maximal with respect to the property $A \cap B = 0$. Such submodules B always exist, by virtue of Zorn's Lemma; in fact, any submodule B_0 of C satisfying $A \cap B_0 = 0$ can be enlarged to a relative complement for A. Of course, if A is actually a direct summand of C, say $C = A \oplus B$, then the complementary summand B is a relative complement for A. For example, if F is a field, $C = F \oplus F$, and $A = F \oplus 0$, then for any $x \in F$ the subspace $(x,1)F$ is a relative complement for A in C. In case F is infinite, this provides an example in which A has infinitely many distinct relative complements in C. (See also Exercises 1 and 2 at the end of this section.)

The importance of relative complements is that they can be used to construct essential submodules, as in the following proposition.

Proposition 1.3 Let $A \leqq C$. If B is any relative complement for A in C, then $A \oplus B \leqq_e C$.

Proof: Since $A \cap B = 0$, we have $A + B = A \oplus B$, so that $A \oplus B$ is a submodule of C. Suppose that $M \leqq C$ with $M \cap (A \oplus B) = 0$. Then the sum $(A \oplus B) + M$ is direct, that is, $(A \oplus B) + M = A \oplus B \oplus M$, whence $A \cap (B \oplus M) = 0$. By the maximality of B, we obtain $B \oplus M = B$ and thus $M = 0$. Therefore $A \oplus B \leqq_e C$.□

This result is so basic to all considerations of essential extensions that we restate it for emphasis: *Every submodule of a module C is a direct summand of an essential submodule of C.*

Definition A submodule A of a module C is said to be a *closed submodule* of C if A has no proper essential extensions inside C, that is, if the only solution of the relation $A \leqq_e B \leqq C$ is $B = A$. For example, 0 and C are always closed submodules of C. Also, every direct summand of C is a closed submodule of C (Exercise 3).

Proposition 1.4 If $B \leqq C$, then the following conditions are equivalent:

(a) B is a closed submodule of C.

(b) B is a relative complement for some $A \leqq C$.

(c) If A is any relative complement for B in C, then B is a relative complement for A in C.

(d) If $B \leqq K \leqq_e C$, then $K/B \leqq_e C/B$.

Proof: (a) $\Rightarrow$ (d): If M/B is a submodule of C/B such that $(M/B) \cap (K/B) = 0$, then $M \cap K = B$. Since $K \leqq_e C$, we have $M \cap K \leqq_e M \cap C$, i.e., $B \leqq_e M$. The assumption that B is closed in C gives us $B = M$, and thus $M/B = 0$.

(d) $\Rightarrow$ (c): Since $A \cap B = 0$, B can be enlarged to a relative complement B' for A. By the modular law, $(A \oplus B) \cap B' = B + (A \cap B') = B$, whence $[(A \oplus B)/B] \cap [B'/B] = 0$. According to 1.3, $A \oplus B \leqq_e C$, and then from (d) we obtain $(A \oplus B)/B \leqq_e C/B$. Thus $B'/B = 0$, and so $B = B'$ is a relative complement for A.

(c) $\Rightarrow$ (b) is automatic.

(b) $\Rightarrow$ (a): Suppose that $B \leqq_e B' \leqq C$. Since $(B' \cap A) \cap B = A \cap B = 0$, we have $B' \cap A = 0$, and then the maximality of B implies that $B' = B$. Thus B is closed in C. □

Proposition 1.5 Let $A \leqq B \leqq C$. If A is closed in B and B is closed in C, then A is closed in C.

Proof: Let A' be a relative complement for A in B, and let B' be a relative complement for B in C. According to 1.3, $B \oplus B' \leqq_e C$; hence 1.4 shows that $(B \oplus B')/B \leqq_e C/B$. We now see from 1.1 that $(B \oplus B')/A \leqq_e C/A$, or $[B/A] \oplus [(A \oplus B')/A] \leqq_e C/A$. Using 1.3 and 1.4 again, we obtain $A \oplus A' \leqq_e B$ and then $(A \oplus A')/A \leqq_e B/A$. According to 1.1, it follows that

$$[(A \oplus A')/A] \oplus [(A \oplus B')/A] \leqq_e C/A$$

or $(A \oplus A' \oplus B')/A \leqq_e C/A$.

Now suppose that we have $A \leqq_e K \leqq C$. Since $A \cap (A' \oplus B') = 0$, 1.1

shows that $K \cap (A' \oplus B') = 0$. Using the modular law, we find that $K \cap (A \oplus A' \oplus B') = A$, whence $[K/A] \cap [(A \oplus A' \oplus B')/A] = 0$. Inasmuch as $(A \oplus A' \oplus B')/A \leqq_e C/A$, we obtain $K/A = 0$, so that $K = A$. Therefore A is closed in C.□

We now present an example to show that the intersection of closed submodules of a module need not be closed.

Example 1.6 There exist modules $A,B \leqq C$ such that A and B are closed in C, but $A \cap B$ is not closed in A, B, or C.

Proof: Set $R = \mathbf{Z}$, $C = \mathbf{Z} \oplus (\mathbf{Z}/2\mathbf{Z})$, $A = (1,0)R$, and $B = (1,\bar{1})R$. Since A is a direct summand of C, it must be closed in C. Observing that $C = B \oplus (0,\bar{1})R$, we see that B is closed in C also. Note that $A \cap B = (2,0)R$. As observed in 1.2, $A \cap B \leqq_e A$ and $A \cap B \leqq_e B$, whence $A \cap B$ is not closed in A, B, or C.□

In general, the essential extensions and submodules of a module can extend infinitely far in either direction. For example, the modules $2^{-n}\mathbf{Z}$ ($n = 1,2,\ldots$) form an infinite ascending sequence of essential extensions of $\mathbf{Z}$, while the ideals $2^n\mathbf{Z}$ ($n = 1,2,\ldots$) form an infinite descending sequence of essential submodules of $\mathbf{Z}$. Both sequences, however, are bounded: The first is bounded above by $\mathbf{Q}$, while the second is bounded below by 0. For any module A, there is a "largest" essential extension of A; this is the injective hull of A, which we study in the next section. On the other hand, there may not be a smallest essential submodule of A (as happens, for example, in $\mathbf{Z}$), but the intersection of all essential submodules of A gives a submodule of A which has no proper essential submodules. This submodule is the socle of A, which we study in Section C.

Exercises

1. Find an example of a module C with a submodule A such that A has relative complements B,B' in C with $A \oplus B = C$, $A \oplus B' < C$.
2. Find an example of a module C with a submodule A such that A has two nonisomorphic relative complements in C.
3. Show that any direct summand of a module C is a closed submodule of C, without using 1.4.
4. If R is a prime ring, show that all nonzero two-sided ideals of R are essential as right ideals. In particular, all nonzero ideals in a commutative integral domain are essential.
5. Find infinite sequences $\{A_n\}$ and $\{B_n\}$ of modules such that $A_n \leqq_e B_n$ for each n, but $\Pi A_n \nleqq_e \Pi B_n$.
6. Find a sequence $\{A_n\}$ of modules such that $\oplus A_n \nleqq_e \Pi A_n$.

7. Show that a module A is noetherian if and only if all essential submodules of A are finitely generated.

8. Prove that a module A_R is flat if and only if the induced map $A \otimes_R J \to A \otimes_R R$ is injective for all essential left ideals J of R.

9. Prove that a module A_R is injective if and only if the induced map $\mathrm{Hom}_R(R,A) \to \mathrm{Hom}_R(J,A)$ is surjective for all essential right ideals J of R.

10. Let R be a commutative ring. Prove that R is semiprime if and only if all finite products of essential ideals of R are essential.

11. If N is a nilpotent two-sided ideal of R, show that its *left* annihilator is an essential *right* ideal of R.

12. Let R be a ring containing exactly one maximal two-sided ideal M. Show that $M_R \leqq_e R_R$ if and only if $M \neq 0$.

13. If $A \leqq C$, prove that C has a closed submodule B such that $A \leqq_e B$.

14. Find an example of a module C with a closed submodule A such that C/A is not isomorphic to any submodule of C.

15. Let $\{A_\alpha\}$ and $\{B_\alpha\}$ be collections of modules such that A_α is a closed submodule of B_α for each α. Prove that $\bigoplus A_\alpha$ must be closed in $\bigoplus B_\alpha$.

16. Let $A \leqq B \leqq C$. If B is closed in C, prove that B/A is closed in C/A. Show that the converse is false.

17. If A is closed in C and $N \leqq_e C$, prove that $A \cap N$ is closed in N.

18. Show that any module has an essential submodule which is a direct sum of cyclic submodules.

19. Let $A \leqq B \leqq C$, and let A' be a relative complement for A in C. Prove that there is some $B' \leqq A'$ such that B' is a relative complement for B in C.

20. The concept of essential extensions dualizes as follows: A submodule A of a module C is called a *small submodule* (or a *superfluous submodule*) of C provided $A + M$ is a proper submodule of C whenever M is a proper submodule of C. Given $A \leqq B \leqq C$, show that B is small in C if and only if A is small in C and B/A is small in C/A.

21. Let $A \leqq B \leqq C$ and $A' \leqq B' \leqq C$. If B/A is small in C/A and B'/A' is small in C/A', show that $(B + B')/(A + A')$ is small in $C/(A + A')$.

22. Let $f : C \to B$. If A is a small submodule of C, show that fA is a small submodule of B.

23. If A_i is a small submodule of B_i for $i = 1, \ldots, n$, prove that $A_1 \oplus \cdots \oplus A_n$ is a small submodule of $B_1 \oplus \cdots \oplus B_n$.

24. Find an example of a module which has a nonzero small essential submodule.

B. Injective Hulls

Proposition 1.7 Let $f : A \to E$ be a monomorphism, where E is injective. Whenever $A \leqq_e B$, f extends to a monomorphism $f' : B \to E$.

Proof: Since E is injective, f must extend to a map $f' : B \to E$. However, $A \cap (\ker f') = \ker f = 0$, and so $\ker f' = 0$.□

Proposition 1.7 shows that the injective module E contains isomorphic copies of all essential extensions of A. Thus E is in some sense an upper bound for essential extensions of A, which suggests that we look inside E for a maximal essential extension of A. We are looking for a supremum (in some sense) of essential extensions of A; by analogy with the real number system, we might find it as an infimum (in some sense) of upper bounds for all essential extensions of A. Thus we should also consider looking for a minimal injective extension of A. As we prove shortly, A does always have a maximal essential extension as well as a minimal injective extension, and they coincide.

Definition An essential monomorphism $f : A \to B$ is called a *proper essential extension of A* whenever fA is a proper (essential) submodule of B.

Proposition 1.8 A module A is injective if and only if A has no proper essential extensions.

Proof: First assume that A is injective, and consider any essential monomorphism $f : A \to B$. Now $fA \cong A$ and so is injective, whence $B = fA \oplus C$ for some C. Since $fA \leqq_e B$, we obtain $C = 0$ and thus $fA = B$. Therefore f is not a proper essential extension of A.

Conversely, if A is not injective, then there exists a module C, containing A, such that A is not a direct summand of C. Choosing a relative complement B for A in C, we see from 1.3 and 1.4 that $(A \oplus B)/B \leqq_e C/B$. Inasmuch as $A \cap B = 0$, the natural map $f : A \to C \to C/B$ is thus an essential extension of A. Since A is not a direct summand of C, $A \oplus B < C$; hence $fA < C/B$. Therefore f is a proper essential extension of A.□

Corollary 1.9 Any closed submodule of an injective module is injective.

Proof: Let A be a closed submodule of an injective module E, and consider any essential monomorphism $f : A \to B$. According to 1.7, the isomorphism $f^{-1} : fA \to A$ extends to a monomorphism $g : B \to E$. Then g is an isomorphism of B onto gB which carries fA onto A; hence we obtain $A \leqq_e gB$. Since A is a closed submodule of E, we must have $gB = A$, whence $B = fA$. Therefore A has no proper essential extensions; hence 1.8 says that A is injective.□

Theorem 1.10 Given any module A, there exists a module E, containing A, such that

(a) E is a maximal essential extension of A in the sense that $A \leqq_e E$, and whenever $A \leqq_e B$, the inclusion map $A \to E$ extends to a monomorphism $B \to E$.

(b) E is a minimal injective extension of A in the sense that E is injective, and any monomorphism $A \to E'$ with E' injective extends to a monomorphism $E \to E'$.

Proof: There exists an injective module F which contains A, and we look for E inside F. According to Zorn's Lemma, we may pick a submodule E of F that is maximal among those submodules of F which are essential extensions of A.

Thus at least inside F we have a maximal essential extension for A. Inasmuch as any essential extension of E inside F is also an essential extension of A, we infer from the maximality of E that E has no proper essential extensions inside F, that is, E is a closed submodule of F. According to 1.9, it follows that E is injective.

We now have $A \leqq_e E$, with E injective, which yields the first parts of (a) and (b). Since E is injective, the second part of (a) follows from 1.7. The second part of (b) also follows from 1.7, because $A \leqq_e E$.□

Definition Any module E satisfying the conditions of 1.10(b) is called an *injective hull* of A (or an *injective envelope* of A). For example, if A is injective, then A is an injective hull of itself. We use the notation $E(A)$ to stand for an injective hull of A. The next proposition shows that $E(A)$ is unique up to isomorphism, but A can certainly have two injective hulls which are distinct as sets. In particular, it can happen that A is a submodule of an injective module F in such a way that F has two distinct submodules which are injective hulls of A. (See Exercise 1.) Thus we use the symbol $E(A)$ when any injective hull of A will suit our purposes, and we use the statement "$B = E(A)$" only as a shorthand for the statement "B is an injective hull of A."

Proposition 1.11 Let A be a module.

(a) If E is an injective module containing A, then E is an injective hull of A if and only if $A \leqq_e E$.

(b) If F is any injective module containing A, then F has at least one submodule which is an injective hull of A.

(c) If E and E' are both injective hulls of A, then the identity map on A extends to an isomorphism of E onto E'.

Proof: (a) If $A \leqq_e E$, then it is clear from 1.7 that E is an injective hull of A. Conversely, assume that E is an injective hull of A. According to 1.10, A has an injective hull E' such that $A \leqq_e E'$. By definition, the inclusion map $A \to E'$ must extend to a monomorphism $f : E \to E'$. Since f is a monomorphism, we infer that $f^{-1}A = A$; hence it follows from 1.1 that $A \leqq_e E$.

(b) As in the proof of 1.10, F has an injective submodule E such that $A \leqq_e E$.

(c) By definition, the inclusion map $A \to E'$ extends to a monomorphism

$f: E \to E'$. We have $A = fA \leqq fE \leqq E'$, and $A \leqq_e E'$ by (a); hence $fE \leqq_e E'$, that is, f is an essential extension of E. According to 1.8, E has no proper essential extensions, whence $fE = E'$. Therefore f is an isomorphism.□

For any modules A and B, we use 1.11 and 1.1 to see that $A \oplus B \leqq_e E(A) \oplus E(B)$ and then that $E(A \oplus B) = E(A) \oplus E(B)$. Similarly, $E(A_1 \oplus \cdots \oplus A_n) = E(A_1) \oplus \cdots \oplus E(A_n)$ for any finite set of modules $A_1, \ldots, A_n$. However, this may fail for infinite direct sums, as Exercise 2 shows. (See also Exercise 3.)

Given any nonzero $n \in \mathbf{Z}$, we have seen that $n\mathbf{Z} \leqq_e \mathbf{Q}$. Since $\mathbf{Q}_{\mathbf{Z}}$ is injective, 1.11 thus shows that $E(n\mathbf{Z}) = \mathbf{Q}$.

Given any prime number p, we note that $\mathbf{Z}(p^\infty)$ has exactly $p - 1$ elements of order p, namely $\overline{1/p}, \overline{2/p}, \ldots, \overline{(p-1)/p}$. Thus $\mathbf{Z}/p\mathbf{Z}$ is isomorphic to the submodule $A = (\overline{1/p})\mathbf{Z}$ of $\mathbf{Z}(p^\infty)$, and we claim that $\mathbf{Z}(p^\infty) = E(A)$.

Setting $E = \mathbf{Z}(p^\infty)$, we note from the definition that $Ep = E$. Now consider any prime $q \neq p$. Given any $x \in E$, we have $xp^n = 0$ for some n. Since q is relatively prime to p^n, we obtain $aq + bp^n = 1$ for some $a,b \in \mathbf{Z}$, whence $xaq = x$. Thus $Eq = E$ for all primes q, hence E is divisible. Therefore E is an injective $\mathbf{Z}$-module.

Any nonzero element $x \in E$ can be written in the form $\overline{m/p^n}$ such that $n > 0$ and m is relatively prime to p. Then m is relatively prime to p^n; hence we have $am + bp^n = 1$ for some $a,b \in \mathbf{Z}$. As a result, $xa = \overline{1/p^n}$, and so $xap^{n-1} = \overline{1/p}$ is a nonzero element of A. Thus $A \leqq_e E$, hence $E = E(A)$, by 1.11.

Therefore $E(\mathbf{Z}/p\mathbf{Z}) \cong \mathbf{Z}(p^\infty)$. More generally, $E(\mathbf{Z}/p^n\mathbf{Z}) \cong \mathbf{Z}(p^\infty)$ for all $n > 0$ (Exercise 4).

If we choose a particular injective hull $E(A)$ for each module A, then we obtain a function E which maps modules to injective modules. On first glance, it would seem easy to make E into a functor: Given any map $f : A \to B$, the injectivity of $E(B)$ implies that f extends to a map $E(f)$: $E(A) \to E(B)$. However, there is no guarantee that $E(f)$ is unique (see Exercise 5), which means that we must choose a particular extension of f to be $E(f)$. After all these choices, can we hope to have E preserve composition of maps? The answer in general is no, as the next proposition shows. (See also Exercise 6 and Exercise 2.A.24.)

Proposition 1.12 There does not exist a functor F : Mod-$\mathbf{Z} \to$ Mod-$\mathbf{Z}$ such that $F(A) \cong E(A)$ for all $A_{\mathbf{Z}}$.

Proof: Assuming, on the contrary, that there does exist such a functor F, we first note that F preserves zero maps. Any zero map $0 : A \to B$ can always be factored through the zero module, that is, $0 = gf$, where $f : A \to 0$ and $g : 0 \to B$. Inasmuch as F of the zero module is isomorphic to $E(0) = 0$, we see that $F(0) = F(g)F(f)$ factors through the zero module, whence F of the zero map $0 : A \to B$ is the zero map $0 : F(A) \to F(B)$.

Now consider the module $A = \mathbf{Z}/2\mathbf{Z}$, and let $j_i : A \to A \oplus A$, $p_i : A \oplus A \to A$ ($i = 1,2$) be the canonical injections and projections of the direct sum. We know that

$$F(A \oplus A) \cong E(A \oplus A) \cong E(A) \oplus E(A) \cong F(A) \oplus F(A)$$

but we do not know a priori whether the maps $F(j_i),F(p_i)$ must be the injections and projections of this direct sum. However, we can show that this is true in the particular case at hand.

Since $p_i j_i = 1_A$ for each i, we obtain $F(p_i)F(j_i) = 1_{F(A)}$ for each i, from which it follows that each $F(j_i)$ is a monomorphism, and that $F(j_1)F(A) \cap [\ker F(p_1)] = 0$. On the other hand, since $p_1 j_2 = 0$ we get $F(p_1)F(j_2) = 0$ and so $F(j_2)F(A) \leqq \ker F(p_1)$. Therefore $F(j_1)F(A) \cap F(j_2)F(A) = 0$; hence the submodule $B = F(j_1)F(A) + F(j_2)F(A)$ of $F(A \oplus A)$ must be isomorphic to $F(A) \oplus F(A)$.

Inasmuch as $F(A \oplus A) \cong E(A \oplus A)$, $F(A \oplus A)$ must have an essential submodule C which is isomorphic to $A \oplus A$. Note that C has exactly four elements, namely 0 and three elements of order 2. As we have just observed above, $E(A) \cong \mathbf{Z}(2^\infty)$, which has exactly one element of order 2. Thus B and $F(A \oplus A)$, which are isomorphic to $E(A) \oplus E(A)$, each have exactly three elements of order 2. As a result, we see that $C \leqq B$, whence $B \leqq_e F(A \oplus A)$. Now $B \cong F(A) \oplus F(A)$, which has no proper essential extensions, by 1.8. Therefore $B = F(A \oplus A)$, that is,

$$F(A \oplus A) = F(j_1)F(A) + F(j_2)F(A)$$

Since $F(p_i)F(j_i) = 1_{F(A)}$ for each i, and since $F(p_i)F(j_k) = 0$ for $i \neq k$, we now compute that

$$F(j_1)F(p_1) + F(j_2)F(p_2) = 1_{F(A \oplus A)}$$

Therefore the maps $F(j_i),F(p_i)$ are indeed the canonical injections and projections for the decomposition $F(A \oplus A) \cong F(A) \oplus F(A)$.

Finally, let $f : A \to A \oplus A$ and $g : A \oplus A \to A$ be the diagonal and sum maps, that is, $f(a) = (a,a)$ and $g(a,a') = a + a'$. Then $p_i f = g j_i = 1_A$ for each i, whence $F(p_i)F(f) = F(g)F(j_i) = 1_{F(A)}$ for each i. Thus

$$F(g)F(f) = F(g)[F(j_1)F(p_1) + F(j_2)F(p_2)]F(f) = 1_{F(A)} + 1_{F(A)}$$

Since $F(A)$ has elements of order greater than 2, it follows that $F(g)F(f) \neq 0$. On the other hand, we have $gf = 0$ and thus $F(gf) = 0$, which is a contradiction. □

We close this section with the following application of the use of injective hulls.

Theorem 1.13 If A,B are injective modules such that A is isomorphic to a submodule of B and B is isomorphic to a submodule of A, then $A \cong B$.

Proof: By assumption, B is isomorphic to a submodule C of A, and there exists a monomorphism $f : A \to C$. Inasmuch as C is injective, we have $A = C \oplus H$ for some H, and we claim that $\{H, fH, f^2H, \ldots\}$ is an independent family of submodules of A.

Since $H \cap fH \leqq H \cap C = 0$, we see that $\{H, fH\}$ is independent. Now suppose that $\{H, fH, \ldots, f^nH\}$ is independent, for some $n > 0$. Since f is a monomorphism, it follows that $\{fH, f^2H, \ldots, f^{n+1}H\}$ is independent. We also have

$$H \cap (fH \oplus f^2H \oplus \cdots \oplus f^{n+1}H) \leqq H \cap fA \leqq H \cap C = 0$$

and consequently $\{H, fH, \ldots, f^{n+1}H\}$ is independent. Therefore the induction works, and consequently $\{H, fH, f^2H, \ldots\}$ is independent, as claimed.

Now A has a submodule $P = H \oplus fH \oplus f^2H \oplus \cdots$, and we note that $P = H \oplus fP$. Inasmuch as fP is a submodule of the injective module C, 1.11 says that C must contain an injective hull Q for fP. Observing that $P = H \oplus fP \leqq_e H \oplus Q$, we use 1.11 again to see that $H \oplus Q$ is an injective hull for P. Since $P \cong fP$, one more application of 1.11 shows that $H \oplus Q \cong Q$. Finally, $C = Q \oplus K$ for some K, whence

$$A = H \oplus C = H \oplus Q \oplus K \cong Q \oplus K = C \cong B. \square$$

As Exercise 7 shows, 1.13 does not hold for noninjective modules in general. (See also Exercise 8.)

Exercises

1. Find an example of an injective module E with submodules A,B,B' such that B and B' are injective hulls of A, but $B \neq B'$. (*Hint*: Example 1.2.)
2. Let $F_1, F_2, \ldots$ be fields, and set $R = \Pi F_n$. Find ideals $A_1, A_2, \ldots$ of R such that $E(\oplus A_n) \ncong \oplus E(A_n)$.
3. Prove that R is right noetherian if and only if $E(\oplus A_\alpha) \cong \oplus E(A_\alpha)$ for all collections $\{A_\alpha\}$ of right R-modules.
4. If p is a prime number and n is a positive integer, show that $E(\mathbf{Z}/p^n\mathbf{Z}) \cong \mathbf{Z}(p^\infty)$.
5. Find an example of a module A such that the map $0 : A \to A$ extends two different ways to maps $E(A) \to E(A)$.
6. Let R be a commutative integral domain which is not a field. Prove that there does not exist a functor F : Mod-$R \to$ Mod-R such that $F(A) \cong E(A)$ for all A_R.
7. Set $A_n = \mathbf{Z}/4\mathbf{Z}$ for $n = 1,2, \ldots$, $A = \oplus A_n$, $B = A \oplus (\mathbf{Z}/2\mathbf{Z})$. Show that A (respectively, B) is isomorphic to a submodule of B (respectively, A), but that $A \ncong B$.

8. If A (respectively, B) is isomorphic to a submodule of B (respectively, A), show that $E(A) \cong E(B)$.
9. Find examples of modules $A \leqq B$ such that $E(B) \ncong E(A) \oplus E(B/A)$.
10. If A is any torsion-free **Z**-module, show that $E(A) \cong A \otimes_{\mathbf{Z}} \mathbf{Q}$.
11. Let R be a commutative integral domain. Given any torsion R-module A, prove that $E(A)$ is a torsion module.
12. Let R be a commutative integral domain. If E is any injective R-module, prove that the torsion submodule of E is a direct summand of E.
13. Prove that $E(A) \cong E(B)$ if and only if there exist $A' \leqq_e A$ and $B' \leqq_e B$ such that $A' \cong B'$.
14. Find an example of a module A such that $\mathrm{End}_R(A)$ is not isomorphic to any subring of $\mathrm{End}_R(E(A))$.
15. Let R be a commutative noetherian ring, and let J be any ideal of R. Given any $x \in E(R/J)$, prove that $xJ^n = 0$ for some n.
16. Let R be a commutative noetherian ring. Given any indecomposable injective R-module E, prove that there exists a unique prime ideal P in R such that $E \cong E(R/P)$.
17. Let R be a commutative noetherian ring, and let P,Q be prime ideals of R. Prove that $P \leqq Q$ if and only if $\mathrm{Hom}_R(E(R/P),E(R/Q)) \neq 0$.
18. Let $F_1,F_2,\ldots$ be fields, $R = \Pi F_n$, $I = \oplus F_n$. Prove that $E(R/I)$ has no nonzero indecomposable injective submodules.
19. Prove that a right R-module A is zero if and only if $\mathrm{Hom}_R(A,E(R/M)) = 0$ for all maximal right ideals M of R.
20. If R has a unique maximal right ideal M, prove that $E(R/M)$ is a faithful right R-module.
21. If R is a commutative noetherian ring, use Exercises 15 and 19 to prove that $\cap J(R)^n = 0$.
22. Let I be a two-sided ideal of R, and let A be any right (R/I)-module. Show that $E(A_{R/I}) = \{x \in E(A_R) \mid xI = 0\}$.

C. Semisimple Modules

Definition For any module A, the sum of all simple submodules of A is called the *socle* of A, denoted $\mathrm{soc}(A)$. [Since the sum of an empty family of submodules of A is by convention equal to 0, we have $\mathrm{soc}(A) = 0$ in case A has no simple submodules.] The module A is said to be *semisimple* (or *completely reducible*) provided $\mathrm{soc}(A) = A$, i.e., provided A is the sum of all its simple submodules. Note that according to these definitions, 0 is a semisimple module.

We note that any homomorphism $f : A \to B$ must carry $\mathrm{soc}(A)$ into $\mathrm{soc}(B)$. For if S is any simple submodule of A, fS is either 0 or simple, whence $fS \leqq \mathrm{soc}(B)$. Therefore $f(\mathrm{soc}(A)) \leqq \mathrm{soc}(B)$. In particular, $f(\mathrm{soc}(A)) \leqq \mathrm{soc}(A)$ for all $f \in \mathrm{End}_R(A)$, so that $\mathrm{soc}(A)$ is a fully invariant submodule of A. For the case $A = R_R$, this means that $\mathrm{soc}(R_R)$ is a two-sided ideal of R. Likewise,

$\mathrm{soc}({}_RR)$ is a two-sided ideal of R, but in general $\mathrm{soc}(R_R) \neq \mathrm{soc}({}_RR)$. (See Exercise 1.)

Proposition 1.14 For any module A, $\mathrm{soc}(A)$ is a direct sum of simple modules. In particular, every semisimple module is a direct sum of simple modules.

Proof: The case $\mathrm{soc}(A) = 0$ is covered by the usual convention that the direct sum of an empty family of modules is 0; thus 0 is the direct sum of an empty family of simple modules.

Now assume that $\mathrm{soc}(A) \neq 0$, and let $\mathscr{S}$ denote the collection of all nonempty independent families of simple submodules of A. Inasmuch as $\mathrm{soc}(A) \neq 0$, A has at least one simple submodule S, whence $\{S\} \in \mathscr{S}$ and so $\mathscr{S}$ is nonempty. According to Zorn's Lemma, $\mathscr{S}$ must contain a maximal family $\{S_\alpha\}$, and we claim that $\mathrm{soc}(A) = \oplus S_\alpha$. Since $\oplus S_\alpha$ is a sum of some of the simple submodules of A, we have $\oplus S_\alpha \leqq \mathrm{soc}(A)$. Thus if $\mathrm{soc}(A) \neq \oplus S_\alpha$, A must have a simple submodule S such that $S \nleqq \oplus S_\alpha$. Now $S \cap (\oplus S_\alpha) \neq S$; hence we obtain $S \cap (\oplus S_\alpha) = 0$. It follows that $\{S_\alpha\} \cup \{S\}$ is independent, which contradicts the maximality of $\{S_\alpha\}$. Therefore $\mathrm{soc}(A) = \oplus S_\alpha$, as claimed.□

For use in the next two propositions, we recall the following fact about internal direct sums: *If $A \leqq B \leqq C$ and A is a direct summand of C, then A is also a direct summand of B.* (See Exercise 2.)

Proposition 1.15 For any module C, the following conditions are equivalent:

(a) C is semisimple.

(b) C has no proper essential submodules.

(c) Every submodule of C is a direct summand of C.

Proof: (a) $\Rightarrow$ (b): If A is a proper submodule of C, then C must have a simple submodule M such that $M \nleqq A$. Then $M \cap A \neq M$; hence $M \cap A = 0$. Thus $A \nleqq_e C$.

(b) $\Rightarrow$ (c): Given any $A \leqq C$, let B be a relative complement for A in C. According to 1.3, $A \oplus B \leqq_e C$, whence (b) says that $A \oplus B = C$.

(c) $\Rightarrow$ (a): In view of (c), we have $C = \mathrm{soc}(C) \oplus A$ for some A, and our task is to show that $A = 0$. If not, choose a nonzero element $x \in A$ and set $J = \{r \in R \mid xr = 0\}$, so that $R/J \cong xR$. Since J is a proper right ideal of R, it must be contained in a maximal right ideal M, and we observe that $xR/xM \cong R/M$, which is a simple module. According to (c), xM is a direct summand of C, and thus of xR. Consequently, $xR = B \oplus xM$ for some B. Now $B \cong xR/xM$ and so B is simple, whence $B \leqq \mathrm{soc}(C)$. However, we also have $B \leqq A$, which is impossible. Therefore $A = 0$, and so $\mathrm{soc}(C) = C$.□

Proposition 1.16 Let $A \leqq C$. Then A is an intersection of essential submodules of C if and only if $\mathrm{soc}(C) \leqq A$. In particular, $\mathrm{soc}(C)$ is the intersection of all the essential submodules of C.

Proof: If S is a simple submodule of C and $B \leqq_e C$, then $S \cap B \neq 0$, from which we obtain $S \cap B = S$, that is, $S \leqq B$. Thus every simple submodule of C is contained in every essential submodule of C. If A is an intersection of essential submodules of C, this shows that $\mathrm{soc}(C) \leqq A$.

Conversely, assume that $\mathrm{soc}(C) \leqq A$, and let K denote the intersection of all those essential submodules of C which contain A. We first claim that any $J \leqq K$ which contains A must be a direct summand of K. According to 1.3, we have $J \oplus B \leqq_e C$ for a suitable B, and since $A \leqq J \oplus B$ we must have $K \leqq J \oplus B$. Now $J \leqq K \leqq J \oplus B$; hence J must be a direct summand of K.

In particular, taking $J = A$, we obtain $K = A \oplus T$ for some T. Given any $M \leqq T$, we have $A \leqq A \oplus M \leqq K$, and so $A \oplus M$ must be a direct summand of K. Then we have $M \leqq T \leqq K$ with M a direct summand of K, from which it follows that M is a direct summand of T. Therefore every submodule of T is a direct summand of T. According to 1.15, T must be semisimple, whence $T \leqq \mathrm{soc}(C)$. Inasmuch as $\mathrm{soc}(C) \leqq A$, we obtain $T = T \cap A = 0$, and so $K = A$. Therefore A is an intersection of essential submodules of C. □

Definition Given any class $\mathfrak{A}$ of modules and any process P for obtaining modules, we say that $\mathfrak{A}$ is *closed under P* provided that any module obtained from modules in $\mathfrak{A}$ by using P must also belong to $\mathfrak{A}$. For example, $\mathfrak{A}$ is closed under submodules if $A \in \mathfrak{A}$ whenever $A \leqq C$ and $C \in \mathfrak{A}$. Other standard situations are closure under factor modules, direct sums, or direct products.

Proposition 1.17 The class of all semisimple right R-modules is closed under submodules, factor modules, and direct sums.

Proof: If C is semisimple and $A \leqq C$, then the natural map $C \to C/A$ must carry $C = \mathrm{soc}(C)$ into $\mathrm{soc}(C/A)$, whence $\mathrm{soc}(C/A) = C/A$. Thus C/A is semisimple. According to 1.15, every submodule of A is a direct summand of C, and thus is a direct summand of A, hence A is semisimple (using 1.15 again). Finally, closure under direct sums is clear from the definitions. (See also Exercise 3.) □

In general, the class of semisimple modules need not be closed under direct products. For example, let $R = \mathbf{Z}$, let $p_1, p_2, \ldots$ be distinct prime numbers, and set $A_n = \mathbf{Z}/p_n\mathbf{Z}$ for all n. Each A_n is a simple R-module, hence semisimple. Since the p_n are distinct primes, $\cap(p_n\mathbf{Z}) = 0$, from which we see that ΠA_n must have a submodule isomorphic to R. Now R is not semisimple; hence it follows from 1.17 that ΠA_n is not semisimple. (See also Exercise 4.)

An obvious question to ask at this point is, when are all modules semisimple? This is answered by the famous Wedderburn-Artin Theorem, one form of which we now state for reference.

Theorem 1.18 The following conditions on a ring R are equivalent:

(a) All right R-modules are semisimple.

(b) R_R is semisimple.

(c) R is semiprime and right artinian.

(d) R is isomorphic to a finite direct product of full matrix rings over division rings.

(e) All right R-modules are projective.

(f) All right R-modules are injective.

(g) All short exact sequences of right R-modules split. □

Since condition (d) is right-left symmetric, the left-hand versions of (a) to (c), (e) to (g) could be added to the list of equivalent conditions in this theorem. In particular, R_R is semisimple if and only if ${}_RR$ is semisimple. Thus we say that R is a *semisimple ring* whenever R_R (or ${}_RR$) is semisimple. The reader should be warned that this terminology is not universal: Some authors use the term "semisimple ring" to stand for a ring whose Jacobson radical is zero. While all semisimple rings have zero Jacobson radical (Exercise 5), the converse is false. For example, $J(\mathbf{Z}) = 0$, but $\mathbf{Z}$ is not a semisimple ring.

Exercises

1. Let F be a field, $R = \begin{pmatrix} F & 0 \\ F[x] & F[x] \end{pmatrix}$. Prove that $\operatorname{soc}(R_R) \leqq_e R_R$ but that $\operatorname{soc}({}_RR) = 0$.
2. Let $A \leqq B \leqq C$. If $C = A \oplus W$, prove that $B = A \oplus (B \cap W)$.
3. If $\{A_\alpha\}$ is any collection of right R-modules, show that $\operatorname{soc}(\bigoplus A_\alpha) = \bigoplus \operatorname{soc}(A_\alpha)$.
4. Prove that the class of all semisimple right R-modules is closed under direct products if and only if $R/J(R)$ is a semisimple ring.
5. If R is semisimple, prove that $J(R) = 0$.
6. Show that a $\mathbf{Z}$-module A is semisimple if and only if every element of A has finite square-free order.
7. If A is a $\mathbf{Z}$-module, show that $\operatorname{soc}(A) \leqq_e A$ if and only if A is torsion.
8. Let $C = \Sigma S_\alpha$, where each S_α is simple. If $A < B \leqq C$ and B/A is simple, prove that $B/A \cong S_\alpha$ for some α.
9. If $A \leqq B$, show that $\operatorname{soc}(A) = A \cap \operatorname{soc}(B)$.
10. If $A \leqq_e B$, show that $\operatorname{soc}(A) = \operatorname{soc}(B)$.
11. Let D be a division ring, V a vector space over D, $R = \operatorname{End}_D(V)$. Prove that $\operatorname{soc}(R_R) = \operatorname{soc}({}_RR) = \{f \in R \mid [fV : D] < \infty\}$.
12. Prove the equivalence of (a), (b), (e), (f), (g) in 1.18.
13. Prove that R is a semisimple ring if and only if all simple right R-modules are projective.

14. Prove that there exists a semisimple module A such that $E(A)$ is faithful.

15. If all semisimple right R-modules are injective, prove that R is right noetherian.

16. If R is semiprime, prove that every finitely generated semisimple right ideal of R is generated by an idempotent.

17. Let R be semiprime, J a two-sided ideal of R. If J_R is finitely generated and semisimple, prove that J is generated by a central idempotent.

18. Let R be semiprime, $e = e^2 \in R$. Prove that $(eR)_R$ is semisimple if and only if ${}_R(Re)$ is semisimple, if and only if eRe is a semisimple ring.

19. If R is semiprime, show that $\mathrm{soc}(R_R) = \mathrm{soc}({}_RR)$.

20. Let R be semiprime. Prove that $\mathrm{soc}(R_R) \leqq_e R_R$ if and only if $\mathrm{soc}({}_RR) \leqq_e {}_RR$.

21. Prove that (b) $\Leftrightarrow$ (c) in 1.18.

22. Dualize 1.16 as follows. The *radical* of a module C is the intersection of all maximal submodules of C and is denoted $\mathrm{rad}(C)$. [If C has no maximal submodules, then $\mathrm{rad}(C) = C$.] Prove that $\mathrm{rad}(C)$ is the sum of all small submodules of C. (See Exercise 1.A.20.) Moreover, show that an element $x \in C$ belongs to $\mathrm{rad}(C)$ if and only if xR is a small submodule of C.

D. The Singular Submodule

Definition We shall use $\mathscr{S}(R)$ to stand for the set of all essential right ideals of the ring R. Also, if I is a right ideal of R and $r \in R$, we use $r^{-1}I$ to denote the right ideal $\{x \in R \mid rx \in I\}$. (Note that if r is invertible in R, then this definition of $r^{-1}I$ coincides with the product of r^{-1} and I.)

Proposition 1.19 (a) $R \in \mathscr{S}(R)$.

(b) If $I \leqq J \leqq R_R$ and $I \in \mathscr{S}(R)$, then $J \in \mathscr{S}(R)$.

(c) If $I,J \in \mathscr{S}(R)$, then $I \cap J \in \mathscr{S}(R)$.

(d) If $I \in \mathscr{S}(R)$ and $r \in R$, then $r^{-1}I \in \mathscr{S}(R)$.

Proof: (a) is obvious, and (b) to (d) follow directly from 1.1.□

Definition Given any right R-module A, we set

$$Z(A) = \{x \in A \mid xI = 0 \text{ for some } I \in \mathscr{S}(R)\}$$

Equivalently, $Z(A)$ is the set of those $x \in A$ for which the right ideal $\{r \in R \mid xr = 0\}$ belongs to $\mathscr{S}(R)$. When the ring R needs to be emphasized, we write $Z_R(A)$ for $Z(A)$.

Since $R \in \mathscr{S}(R)$, we always have $0 \in Z(A)$. If $x,y \in Z(A)$, then $xI = yJ = 0$ for suitable $I,J \in \mathscr{S}(R)$. Since $I \cap J \in \mathscr{S}(R)$ and $(x - y)(I \cap J) = 0$, we get $x - y \in Z(A)$. Also, given any $r \in R$, we have $r^{-1}I \in \mathscr{S}(R)$ and $(xr)(r^{-1}I) \leqq xI = 0$, whence $xr \in Z(A)$. Thus $Z(A)$ is a submodule of A: It is called the *singular submodule* of A.

In a similar fashion, we define the singular submodule of any left R-module B: $Z(B)$ is the set of those $x \in B$ such that $Jx = 0$ for some $J \leqq_e {}_RR$.

Actually, $Z(-)$ defines a functor from Mod-$R \to$ Mod-R. Given any map $f : A \to B$ in Mod-R, it follows directly from our definitions that $f(Z(A)) \leqq Z(B)$; hence we define $Z(f) : Z(A) \to Z(B)$ to be the restriction of f to $Z(A)$. In particular, for any module A we have $f(Z(A)) \leqq Z(A)$ for all $f \in \text{End}_R(A)$, so that $Z(A)$ is a fully invariant submodule of A.

Considering R as a module, we see that $Z(R_R)$ is thus a two-sided ideal of R. The ideal $Z(R_R)$ is known as the *right singular ideal* of R, and is denoted $Z_r(R)$. Likewise, we have the *left singular ideal* $Z_l(R)$, which is the singular submodule of the left module ${}_RR$. In general, $Z_r(R) \neq Z_l(R)$, as shown in Exercise 1.

A module A is called a *singular module* provided $Z(A) = A$. At the other extreme, we say that A is a *nonsingular module* provided $Z(A) = 0$. Thus the ring R is a nonsingular right module if and only if $Z_r(R) = 0$, and in this event R is called a *right nonsingular ring*. (On the other hand, R can never be a singular module unless $R = 0$—see Exercise 2.) Likewise, we say that R is a *left nonsingular ring* if $Z_l(R) = 0$. As Exercise 1 shows, right and left nonsingularity are not equivalent.

We now illustrate these concepts in a simple case, namely $R = \mathbf{Z}$. (The details are similar for any commutative integral domain.) As we noted in Section A, all nonzero ideals of $\mathbf{Z}$ are essential in $\mathbf{Z}$; hence $\mathscr{S}(\mathbf{Z})$ is just the set of all nonzero ideals of $\mathbf{Z}$. Given a $\mathbf{Z}$-module A and an element $x \in A$, we thus have $x \in Z(A)$ if and only if $x(n\mathbf{Z}) = 0$ for some positive integer n, i.e., if and only if x has finite order. Therefore $Z(A)$ is just the torsion subgroup of A. It follows that A is singular if and only if it is a torsion group, and that A is nonsingular if and only if it is a torsion-free group. In particular, $\mathbf{Z}_{\mathbf{Z}}$ is nonsingular; hence $\mathbf{Z}$ is a nonsingular ring.

Proposition 1.20 (a) A module C is nonsingular if and only if $\text{Hom}_R(A,C) = 0$ for all singular modules A.

(b) A module C is singular if and only if there exists a short exact sequence $0 \to A \xrightarrow{f} B \xrightarrow{g} C \to 0$ such that f is an essential monomorphism.

Proof: (a) If A is singular, C is nonsingular, and $f : A \to C$, then $fA = f(Z(A)) \leqq Z(C) = 0$ and $f = 0$. Thus $\text{Hom}_R(A,C) = 0$ whenever A is singular and C is nonsingular.

Conversely, if $\text{Hom}_R(A,C) = 0$ for all singular modules A, then in particular $\text{Hom}_R(Z(C),C) = 0$. Now the inclusion map $Z(C) \to C$ is zero, hence $Z(C) = 0$.

(b) First assume that we have such an exact sequence. Given any $b \in B$, we have a map $k : R \to B$ defined by $k(r) = br$. According to 1.1, $k^{-1}(fA) \leqq_e R_R$, that is, the right ideal $I = \{r \in R \mid br \in fA\}$ belongs to $\mathscr{S}(R)$. Now $bI \leqq fA = \ker g$; hence $(gb)I = 0$ and so $gb \in Z(C)$. Since g is an epimorphism, we thus obtain $Z(C) = C$.

Conversely, assume that C is singular, and choose a short exact sequence $0 \to A \xrightarrow{\leq} B \xrightarrow{g} C \to 0$ such that B is free. If $\{b_\alpha\}$ is a basis for B, then for each α we have $(gb_\alpha)I_\alpha = 0$ for some $I_\alpha \in \mathscr{S}(R)$; hence $b_\alpha I_\alpha \leqq A$. Since $I_\alpha \leqq_e R_R$ for all α, we get $b_\alpha I_\alpha \leqq_e b_\alpha R$ for all α; hence 1.1 says that $\oplus b_\alpha I_\alpha \leqq_e \oplus b_\alpha R = B$. Inasmuch as $\oplus b_\alpha I_\alpha \leqq A$, we obtain $A \leqq_e B$, and thus the inclusion map $A \to B$ is an essential monomorphism.□

In particular, 1.20 shows that B/A is singular whenever $A \leqq_e B$. Thus, for example, $E(A)/A$ is always singular. The converse of this can easily fail; for example, let $B = \mathbf{Z}/2\mathbf{Z}$ and $A = 0$. As we have noted above, B/A is a singular $\mathbf{Z}$-module, and yet $A \nleqq_e B$. There are, however, two special cases in which this converse does work: One is given by the next proposition, and the other is the case $B = R_R$. Namely, if I is a right ideal of R such that R/I is singular, then we must have $\bar{1}J = 0$ for some $J \in \mathscr{S}(R)$. Then $J \leqq I$ and so $I \in \mathscr{S}(R)$, that is, $I \leqq_e R_R$. Since we already know that R/I is singular whenever $I \leqq_e R_R$, we conclude that

$$\mathscr{S}(R) = \{I \leqq R_R \mid R/I \text{ is singular}\}$$

Proposition 1.21 Let B be nonsingular, and let $A \leqq B$. Then B/A is singular if and only if $A \leqq_e B$.

Proof: If B/A is singular and x is a nonzero element of B, then $\bar{x}I = 0$ for some $I \in \mathscr{S}(R)$, that is, $xI \leqq A$. Inasmuch as B is nonsingular, we have $xI \neq 0$ and thus $xR \cap A \neq 0$. Therefore $A \leqq_e B$.□

Definition Let $\mathfrak{A}$ be any class of modules. Following the procedure indicated in the last section, we say that $\mathfrak{A}$ is *closed under essential extensions* provided $B \in \mathfrak{A}$ whenever $A \in \mathfrak{A}$ and $A \leqq_e B$. There is another type of extension which we must also keep in mind in this context. Namely, given any short exact sequence $0 \to C \to B \to A \to 0$ of modules, the middle term B is called a *module extension of C by A* (or just an *extension of C by A*). Thus we say that $\mathfrak{A}$ is *closed under module extensions* provided $B \in \mathfrak{A}$ whenever B is an extension of a module $C \in \mathfrak{A}$ by a module $A \in \mathfrak{A}$.

Proposition 1.22 (a) The class of all nonsingular right R-modules is closed under submodules, direct products, essential extensions, and module extensions.

(b) The class of all singular right R-modules is closed under submodules, factor modules, and direct sums.

Proof: (a) Whenever $A \leqq B$, we obviously have $Z(A) = A \cap Z(B)$. Thus if B is nonsingular, then so is A. On the other hand, if A is nonsingular and $A \leqq_e B$, then since $A \cap Z(B) = Z(A) = 0$ we must have $Z(B) = 0$ as well. Therefore submodules and essential extensions of nonsingular modules are nonsingular.

If $\{C_\alpha\}$ is any collection of nonsingular modules and A is singular, then $\mathrm{Hom}_R(A,C_\alpha) = 0$ for all α by 1.20, whence $\mathrm{Hom}_R(A,\Pi C_\alpha) = 0$. Now 1.20 says that ΠC_α is nonsingular.

Suppose that $0 \to C \to B \to A \to 0$ is an exact sequence of modules with C,A nonsingular. According to 1.20, we have $\mathrm{Hom}_R(M,C) = 0$ and $\mathrm{Hom}_R(M,A) = 0$ for any singular module M. By exactness of the sequence

$$0 \to \mathrm{Hom}_R(M,C) \to \mathrm{Hom}_R(M,B) \to \mathrm{Hom}_R(M,A)$$

we obtain $\mathrm{Hom}_R(M,B) = 0$ as well, and then 1.20 shows that B is nonsingular.

(b) If $A \leqq B$ and B is singular, then $Z(A) = A \cap Z(B) = A$ and so A is singular. Also, the map $B \to B/A$ must carry $Z(B)$ into $Z(B/A)$; hence $B/A = Z(B)/A \leqq Z(B/A)$ and so B/A is singular.

If $\{C_\alpha\}$ is any collection of singular modules, then for each α 1.20 gives us a short exact sequence $0 \to A_\alpha \to B_\alpha \to C_\alpha \to 0$ such that $A_\alpha \to B_\alpha$ is an essential monomorphism. Now $0 \to \oplus A_\alpha \to \oplus B_\alpha \to \oplus C_\alpha \to 0$ is exact too, and 1.1 says that $\oplus A_\alpha \to \oplus B_\alpha$ is an essential monomorphism; hence by 1.20 we see that $\oplus C_\alpha$ is singular.□

If 1.22(b) seems somewhat incomplete compared to 1.22(a), it is: Without some further hypotheses, we cannot conclude that the singular modules are closed under either module extensions or essential extensions. For example, let $R = \mathbf{Z}/4\mathbf{Z}$, and note that R has exactly three ideals: 0, $2R$, R. Since every nonzero ideal of R contains $2R$, we obtain $\mathscr{S}(R) = \{2R,R\}$. Now $(2R)(2R) = 0$; hence $2R \leqq Z(R)$. Since $1 \notin Z(R)$, it follows that $Z(R) = 2R$. Now $2R$ is a singular R-module, and since $R/2R \cong 2R$, $R/2R$ is singular as well. Thus R is an extension of the singular module $2R$ by the singular module $R/2R$, yet R is not singular. We also note that R is an essential extension of the singular module $2R$. Therefore the class of all singular R-modules is not closed under either module extensions or essential extensions. We can rectify this situation, however, by restricting ourselves to nonsingular rings, as the next proposition shows. (We can also improve the situation by enlarging the class of singular modules somewhat, as in Exercises 20 and 21.)

Proposition 1.23 Assume that $Z_r(R) = 0$.

(a) $Z(A/Z(A)) = 0$ for all right R-modules A.

(b) A right R-module A is singular if and only if $\mathrm{Hom}_R(A,C) = 0$ for all nonsingular right R-modules C.

(c) The class of all singular right R-modules is closed under module extensions and essential extensions.

(d) $\mathscr{S}(R)$ is closed under finite products.

Proof: (a) The module A has a submodule C containing $Z(A)$ such that $C/Z(A) = Z(A/Z(A))$, and we first claim that $Z(A) \leqq_e C$. If $M \leqq C$ and $M \cap Z(A) = 0$, then M is clearly nonsingular. On the other hand, since $M \cap Z(A) = 0$ we get a monomorphism $M \to C \to C/Z(A)$ and since $C/Z(A)$ is singular, it follows that M must be singular. The only way for M to be nonsingular and also singular is $M = 0$, and thus $Z(A) \leqq_e C$, as claimed.

If $Z(A/Z(A)) \neq 0$, then $Z(A) < C$ and we can pick an element $x \in C - Z(A)$. Setting $J = \{r \in R \mid xr = 0\}$, we must have $J \notin \mathscr{S}(R)$, because $x \notin Z(A)$. Thus R has a nonzero right ideal K for which $K \cap J = 0$, and we observe from this that the natural map $K \to xK$ is an isomorphism. Inasmuch as $Z_r(R) = 0$, we have $Z(K) = 0$, and so $Z(xK) = 0$. But then $xK \cap Z(A) = 0$, which contradicts the fact that $Z(A) \leqq_e C$. Therefore $Z(A/Z(A)) = 0$.

(b) If A is singular, then we know from 1.20 that $\mathrm{Hom}_R(A,C) = 0$ for all nonsingular modules C. Conversely, assume that $\mathrm{Hom}_R(A,C) = 0$ for all nonsingular modules C. According to (a), $A/Z(A)$ is nonsingular; hence the natural map $A \to A/Z(A)$ must be zero, and so $Z(A) = A$.

(c) Let $0 \to C \to B \to A \to 0$ be an exact sequence of right R-modules with C,A singular. According to (b), we have $\mathrm{Hom}_R(C,M) = 0$ and $\mathrm{Hom}_R(A,M) = 0$ for any nonsingular module M. By exactness of the sequence

$$0 \to \mathrm{Hom}_R(A,M) \to \mathrm{Hom}_R(B,M) \to \mathrm{Hom}_R(C,M)$$

we obtain $\mathrm{Hom}_R(B,M) = 0$, and then (b) shows that B is singular. Thus the singular right R-modules are closed under module extensions. If A is a singular right R-module and $A \leqq_e B$, then B/A is singular by 1.20. Inasmuch as B is an extension of A by B/A, the previous result shows that B must be singular.

(d) If $I,J \in \mathscr{S}(R)$, then R/I is a singular right R-module. Since $(I/IJ)J = 0$, I/IJ is singular also. According to (c), R/IJ is singular, and thus $IJ \in \mathscr{S}(R)$. The result now follows by induction.□

In the category Mod-R, the singular and nonsingular modules can be identified in categorical terms by using 1.20. Thus any Morita-equivalence Mod-R → Mod-S must take singular R-modules to singular S-modules, and nonsingular R-modules to nonsingular S-modules. Using 1.22, we see that $Z_r(R) = 0$ if and only if all projective right R-modules are nonsingular. Inasmuch as we can describe the latter condition in categorical terms, it follows that any ring Morita-equivalent to a right nonsingular ring must also be a right nonsingular ring. In particular, the ring of all $n \times n$ matrices over a right nonsingular ring is again a right nonsingular ring.

Proposition 1.24 If A is any simple right R-module, then A is either singular or projective, but not both.

Proof: Inasmuch as $A \cong R/M$ for some maximal right ideal M of R, we see that A is singular if and only if $M \in \mathscr{S}(R)$. Thus if A is not singular, we must have $K \cap M = 0$ for some nonzero right ideal K of R. Since M is a maximal right ideal, we obtain $K \oplus M = R$, whence A is projective. Conversely, if A is projective we have $K \oplus M = R$ for some right ideal K, whence $M \notin \mathscr{S}(R)$ and so A is not singular.□

Corollary 1.25 Every nonsingular semisimple right R-module is projective.

Proof: According to 1.14, any semisimple right R-module has the form $\oplus S_\alpha$, where each S_α is simple. If $\oplus S_\alpha$ is nonsingular, then every S_α is nonsingular and thus projective, by 1.24. Therefore $\oplus S_\alpha$ is projective.□

Corollary 1.26 Let $J = \text{soc}(R_R)$. If A is any nonsingular right R-module, then $\text{soc}(A) = AJ$.

Proof: For any $x \in A$, xJ is an epimorphic image of J and so is semisimple, whence $xJ \leqq \text{soc}(A)$. Thus $AJ \leqq \text{soc}(A)$. On the other hand, any simple submodule S of A is projective by 1.24, whence $S \cong eR$ for some idempotent $e \in R$. Since eR is simple, we have $eR \leqq J$. There is an isomorphism $f: eR \to S$, from which we obtain $S = f(eR) = (fe)(eR) \leqq AJ$. Thus $\text{soc}(A) \leqq AJ$.□

When constructing examples, and in some other situations, we must be able to tell that the rings involved are nonsingular. Thus we now present some criteria for determining nonsingularity in certain well-behaved situations.

Proposition 1.27 (a) If all principal right ideals of R are projective, then $Z_r(R) = 0$.

(b) Suppose that R is commutative. Then $Z(R) = 0$ if and only if R is semiprime.

(c) Assume that J is a semisimple right ideal of R whose left annihilator is zero. Then $Z_r(R) = 0$, $\text{soc}(R_R) = RJ$, and $\mathscr{S}(R) = \{I \leqq R_R \mid RJ \leqq I\}$.

Proof: (a) For any $x \in R$, we have a split exact sequence $0 \to I \to R \to xR \to 0$, where $I = \{r \in R \mid xr = 0\}$; hence $R_R = I \oplus J$ for some J. If $x \neq 0$, then $I < R$ and $J \neq 0$. Since $I \cap J = 0$, this means that $I \notin \mathscr{S}(R)$ and so $x \notin Z_r(R)$.

(b) Assume that $Z(R) = 0$, and let I be any ideal of R such that $I^2 = 0$. If J is a relative complement for I in R, then by 1.3 we have $I \oplus J \in \mathscr{S}(R)$. Observing that $I(I \oplus J) = 0$, we see that $I \leqq Z(R) = 0$. Therefore R is semiprime.

On the other hand, if $Z(R) \neq 0$ there must be a nonzero element $x \in R$

such that $xI = 0$ for some $I \in \mathcal{S}(R)$. Since $x \neq 0$, we must have $xR \cap I \neq 0$. Noting that $(xR \cap I)^2 \leqq xI = 0$, we conclude that R is not semiprime.

(c) In view of 1.16, we obtain $J \leqq \mathrm{soc}(R_R) \leqq I$ for all $I \in \mathcal{S}(R)$, from which we see that every $I \in \mathcal{S}(R)$ has zero left annihilator. Thus $Z_r(R) = 0$.

Since J is contained in the two-sided ideal $\mathrm{soc}(R_R)$, we must have $RJ \leqq \mathrm{soc}(R_R)$. On the other hand, for any nonzero $x \in R$ we have $xJ \neq 0$ and thus $xR \cap RJ \neq 0$. Then $RJ \in \mathcal{S}(R)$; hence $\mathrm{soc}(R_R) \leqq RJ$, by 1.16. Thus $\mathrm{soc}(R_R) = RJ$. Using 1.16 once again, we find that $RJ \leqq I$ for all $I \in \mathcal{S}(R)$. On the other hand, since $RJ \in \mathcal{S}(R)$, 1.19 shows that every right ideal of R which contains RJ must belong to $\mathcal{S}(R)$. Therefore $\mathcal{S}(R) = \{I \leqq R_R \mid RJ \leqq I\}$.□

In particular, 1.27(a) shows that all right semihereditary rings are right nonsingular. Thus every regular ring is right and left nonsingular. Also, 1.27(a) shows that any direct product (finite or infinite) of integral domains is right and left nonsingular. (See also Exercise 3.) Of course, direct products of commutative integral domains are also covered by 1.27(b).

Proposition 1.28 Let H be a two-sided ideal of R such that $(R/H)_R$ is nonsingular.

(a) $\mathcal{S}(R/H) = \{I/H \mid I \in \mathcal{S}(R) \text{ and } H \leqq I\}$.

(b) For any right (R/H)-module A, $Z_{R/H}(A) = Z_R(A)$.

(c) $Z_r(R/H) = 0$.

Proof: (a) If $I/H \in \mathcal{S}(R/H)$, then I/H must be an essential right R-submodule of R/H; hence 1.1 shows that $I \in \mathcal{S}(R)$. On the other hand, if $I \in \mathcal{S}(R)$ and $H \leqq I$, then $(R/H)/(I/H) \cong R/I$ is a singular right R-module. According to 1.21, I/H is an essential right R-submodule of R/H, whence $I/H \in \mathcal{S}(R/H)$.

(b) and (c) are direct consequences of (a).□

Exercises

1. Let F be a field, and set $R = \begin{pmatrix} F & 0 \\ F[x]/(x^2) & F[x]/(x^2) \end{pmatrix}$. Show that $Z_r(R) = 0$, and that $Z_l(R) \leqq_e {}_RR$.
2. Prove that $Z_r(R)$ contains no nonzero idempotents.
3. Assume that R has no nonzero nilpotent elements. Show that the right (or left) annihilator of any element of R is a two-sided ideal. Use this to prove that $Z_r(R) = Z_l(R) = 0$.
4. Let $Z_r(R) = 0$, and let $A \leqq B$ be right R-modules. Show that B/A is singular if and only if $A + Z(B) \leqq_e B$.
5. If R is commutative, prove that $R/Z(R)$ is a nonsingular ring.
6. Let F be a field, and set $R = \begin{pmatrix} F[x] & 0 \\ F[x,y]/(x,y^2) & F[y]/(y^2) \end{pmatrix}$. Prove that $Z_r(R/Z_r(R)) \neq 0$.

7. If A is a nonsingular right R-module, prove that A has an essential submodule of the form $\oplus A_\alpha$, where each A_α is isomorphic to a right ideal of R.
8. Show that the class of all singular right R-modules is closed under direct products if and only if $\text{soc}(R_R) \in \mathcal{S}(R)$.
9. Let $J = \text{soc}(R_R)$ and assume that $J \in \mathcal{S}(R)$. Prove that $Z_r(R) = 0$ if and only if $J^2 = J$.
10. If R is semiprime and $\text{soc}(R_R) \in \mathcal{S}(R)$, prove that $Z_r(R) = Z_l(R) = 0$.
11. Set $J = \text{soc}(R_R)$. Prove that $J \leqq Z_r(R)$ if and only if $J^2 = 0$, if and only if J is nil.
12. Assume there is a ring decomposition $R = \Pi R_\alpha$. If I is a right ideal of R, show that $I \in \mathcal{S}(R)$ if and only if I contains a right ideal of the form $\oplus I_\alpha$, where each $I_\alpha \in \mathcal{S}(R_\alpha)$. Conclude that $Z_r(R) = \Pi Z_r(R_\alpha)$. Thus any direct product of right nonsingular rings is a right nonsingular ring.
13. If R is a right nonsingular ring and $x_1, \ldots, x_n$ are indeterminates, prove that $R[x_1, \ldots, x_n]$ is a right nonsingular ring.
14. Show that $Z_r(R) \in \mathcal{S}(R)$ if and only if $Z(A) \neq 0$ for all nonzero right R-modules A.
15. Let R be commutative, and let $P(R)$ denote the prime radical of R. Show that $P(R) \leqq_e Z(R)$. Find an example where $P(R) < Z(R)$. (*Hint*: Exercise 12.)
16. Let S be a commutative ring, and let A be a faithful S-module. Make the abelian group $R = S \oplus A$ into a commutative ring by using the multiplication rule $(s,a)(s',a') = (ss',sa' + s'a)$. Show that $Z(R) = T \oplus A$, where $T = \{s \in S \mid Bs = 0 \text{ for some } B \leqq_e A\}$. (The ideal T is called the *tertiary radical* of A.)
17. Use Exercise 16 to construct a commutative ring R such that $Z(R) \nleqq J(R)$.
18. Find an example of a module A such that $Z(A/Z(A)) \neq 0$.
19. Prove that $Z_r(R) = 0$ if and only if $Z(A/Z(A)) = 0$ for all right R-modules A.
20. Given any module A, define a submodule $Z_2(A)$ by the rule $Z_2(A)/Z(A) = Z(A/Z(A))$. Prove that $Z(A/Z_2(A)) = 0$. Show that $Z_2(R_R)$ is a two-sided ideal of R, and that $R/Z_2(R_R)$ is a right nonsingular ring.
21. Say that a module A is Z_2-*torsion* whenever $Z_2(A) = A$. Prove that the class of all Z_2-torsion right R-modules is closed under submodules, factor modules, direct sums, module extensions, and essential extensions.
22. Let p be a prime number, and n an integer greater than 1. If $R = \mathbf{Z}/p^n\mathbf{Z}$, prove that all R-modules are Z_2-torsion.
23. Prove that any ultraproduct of right nonsingular rings is a right nonsingular ring.
24. Let $Z_r(R) = 0$, and set $J = \text{soc}(R_R)$. Prove that ${}_R(R/J)$ is flat.
25. Let R be a semiprime right nonsingular ring, and set $J = \text{soc}(R_R)$. If the global weak dimension of R/J is at most 1, prove that the global weak dimension of R is at most 1.
26. Let $Z_r(R) = 0$, and set $J = \text{soc}(R_R)$. Prove that

$$\text{r.gl.dim.}(R/J) \leqq \text{r.gl.dim.}(R) \leqq 1 + \text{r.gl.dim.}(R/J)$$

27. Let F be a field, V a vector space over F, $Q = \text{End}_F(V)$. If R is the F-subalgebra of Q generated by 1 and $\text{soc}(Q_Q)$, use Exercise 26 to show that R is right and left hereditary.

2

Localization and Maximal Quotient Rings

The first section of this chapter is devoted to the construction and basic properties of a "localization" functor S° similar to the localizations used in commutative ring theory. This construction is carried out for any right nonsingular ring R, and yields a regular self-injective ring $S^\circ R$, along with an exact additive functor S° from right R-modules to right $S^\circ R$-modules. Section B is concerned with one source of such regular self-injective rings—namely, the endomorphism ring of any quasi-injective module (modulo the Jacobson radical). The third section covers a construct similar to the ring $S^\circ R$—the maximal quotient ring, which exists for any ring. (When R is nonsingular, the maximal quotient ring is exactly $S^\circ R$.) Finally, Section D provides an answer to the question of which right and left nonsingular rings have coinciding maximal right and left quotient rings.

A. Localization

In commutative ring theory, the term "localization" refers to a procedure in which the elements of some multiplicatively closed subset S in a commutative ring R become invertible in a new ring $S^{-1}R$. In fact, S^{-1} is an exact additive functor from R-modules to $S^{-1}R$-modules. For any R-module A, the elements of $S^{-1}A$ are "fractions" a/s ($a \in A$, $s \in S$), where $a/s = a'/s'$ if and only if $t(as' - a's) = 0$ for some $t \in S$, and the module structure on $S^{-1}A$ is given by the usual addition and multiplication rules for fractions.

One particular case of this procedure is the model for the "localization" which we construct in this section, namely, when R is a commutative integral domain and S is the set of all nonzero elements of R. In this case $S^{-1}R$ is just the quotient field of R. Here we can give an alternate description of S^{-1} which does not depend on commutativity or on inverting elements. For any R-module A, we factor out the torsion submodule $T(A)$, and then it turns out that $S^{-1}A$ is an injective hull for $A/T(A)$.

For a general ring R, our analog for $T(A)$ is the singular submodule $Z(A)$. In order to make sure that $Z(-)$ behaves as much as possible like $T(-)$ [e.g., we need to have $Z(A/Z(A)) = 0$], we restrict ourselves to the case $Z_r(R) = 0$. Following the analogy, we then construct $E(A/Z(A))$ for each module A. Since $E(-)$ is not in general functorial (see 1.12), there is a question as to whether we obtain a functor from this construction. However, since we are mainly dealing with the nonsingular modules $A/Z(A)$, there is enough uniqueness to give us functoriality, which is the purpose of the following trivial but very useful lemma.

Lemma 2.1 Assume we are given modules $A \leqq B$, and C, such that B/A is singular and C is nonsingular. Then any two homomorphisms from B to C which agree on A must be equal.

Proof: According to 1.20, $\mathrm{Hom}_R(B/A,C) = 0$, hence we obtain an exact sequence $0 \to \mathrm{Hom}_R(B,C) \to \mathrm{Hom}_R(A,C)$.□

Definition Assume that $Z_r(R) = 0$. For each right R-module A, we choose a particular injective hull for $A/Z(A)$ and label it $S^\circ A$. In view of 1.23 and 1.22, we see that $S^\circ A$ is a nonsingular right R-module.

Given any homomorphism $f : A \to B$ of right R-modules, we have seen that $f(Z(A)) \leqq Z(B)$, whence f induces a map $\bar{f} : A/Z(A) \to B/Z(B)$. According to 2.1, $\bar{f}$ extends *uniquely* to a map from $S^\circ A \to S^\circ B$, and we label this map $S^\circ f$.

If A is a right R-module and $f : A \to A$ is the identity map, then $\bar{f}$ is the identity map on $A/Z(A)$; hence by uniqueness $S^\circ f$ must be the identity map on $S^\circ A$. Thus S° preserves identity maps. Also, given $f : A \to B$ and $g : B \to C$, we see that $\overline{gf} = (\bar{g})(\bar{f})$, whence $S^\circ(gf)$ and $(S^\circ g)(S^\circ f)$ are both extensions of $\overline{gf}$. By uniqueness, $S^\circ(gf) = (S^\circ g)(S^\circ f)$, and therefore S° is a functor. Finally, if we have $f,g : A \to B$, then $\overline{f+g} = \bar{f} + \bar{g}$, hence one more application of uniqueness shows that $S^\circ(f + g) = S^\circ f + S^\circ g$. Thus S° is additive. As a consequence, S° preserves finite direct sums.

At this point, we have an additive functor S° : Mod-R → Mod-R. To continue the analogy with the commutative situation, we should expect three more properties: $S^\circ R$ should be a ring, S° should be a functor from R-modules to $S^\circ R$-modules, and S° should be exact. With regard to the first property, we already have a start toward making $S^\circ R$ into a ring. Namely, $R \subseteq S^\circ R$ (because of our assumption that $Z_r(R) = 0$); hence we already have products in $S^\circ R$ of the form xr, where $x \in S^\circ R$ and $r \in R$. Also, since

$S°R$ is a module, it has an abelian group structure. If we can find a ring structure on $S°R$ which uses this addition, and which extends the module multiplication, we shall say that the ring structure is *compatible* with the R-module structure. Similarly, any $S°R$-module structure on some $S°A$ which uses the R-module addition and extends the R-module multiplication is said to be *compatible* with the R-module structure on $S°A$.

Theorem 2.2 Let $Z_r(R) = 0$.

(a) $S°R$ has a unique ring structure compatible with its right R-module structure.

(b) For any right R-module A, $S°A$ has a unique right $S°R$-module structure compatible with its right R-module structure.

(c) $S°$ is an additive functor from Mod-R to Mod-$S°R$.

Proof: (a) Let Q be the ring $\text{End}_R(S°R)$. We have a **Z**-homomorphism $\phi : Q \to S°R$ given by the rule $\phi(f) = f(1)$, and we claim that ϕ is an isomorphism. Given any $f \in \ker \phi$, we have $fR = 0$ and thus $f = 0$, by 2.1. Second, for any $x \in S°R$ there is a map $g : R_R \to S°R$ such that $g(1) = x$, and g must extend to a map $h \in Q$ for which $\phi(h) = x$. Thus ϕ is indeed an isomorphism.

Since we have a **Z**-isomorphism ϕ from the ring Q to the module $S°R$, we can make $S°R$ into a ring by using the given R-module addition together with the multiplication rule $x \cdot y = \phi[(\phi^{-1}x)(\phi^{-1}y)]$. Note that

$$x \cdot y = (\phi^{-1}x)(\phi^{-1}y)(1) = (\phi^{-1}x)(y)$$

For any $x \in S°R$ and $r \in R$, we compute that

$$x \cdot r = (\phi^{-1}x)(r) = [(\phi^{-1}x)(1)]r = xr$$

Therefore the ring structure on $S°R$ is compatible with its right R-module structure. Because of this compatibility, we now write xy for a product $x \cdot y$ in $S°R$.

Now suppose there is another multiplication $* : S°R \times S°R \to S°R$ which, together with the R-module addition, gives $S°R$ a ring structure compatible with its right R-module structure. Given any $x \in S°R$, we define a right R-module homomorphism $f : S°R \to S°R$ according to the rule $fy = x*y - xy$. Observing that $fR = 0$, we use 2.1 to see that $f = 0$. Therefore the ring structure on $S°R$ is unique.

(b) With Q as above, the group $K = \text{Hom}_R(S°R, S°A)$ becomes a right Q-module by using composition of functions as the module multiplication. As in (a), we obtain a **Z**-isomorphism $\psi : K \to S°A$ according to the rule $\psi(f) = f(1)$. This allows us to make $S°A$ into a right $S°R$-module, and we check compatibility and uniqueness as in (a).

(c) In view of (b), $S°$ takes right R-modules to right $S°R$-modules, and we already know that $S°$ preserves identity maps, and composition and sums of maps. All that remains is to show that $S°$ takes any right R-module map $f : A \to B$ into a right $S°R$-module map. Given any $x \in S°A$, we define a right R-homomorphism $g : S°R \to S°B$ according to the rule

$$gy = (S°f)(xy) - [(S°f)(x)]y$$

Clearly $gR = 0$; hence 2.1 says that $g = 0$. Therefore $S°f$ is indeed an $S°R$-homomorphism.□

As far as the properties given in 2.2 are concerned, the functor $S°$ is only one case of a more general procedure. For any ring R, there is a class of functors T : Mod-$R \to$ Mod-R, called "torsion theories," which can be used in the same manner as we have used $Z(-)$. (See, for example, Z_2 in Exercise 1.D.20 and T_r in Exercise 2.C.19.) Each such T induces a functor similar to $S°$, called the "localization functor" associated with T. (See, for example, $S_r°$ in Exercise 2.C.21.) In this generality, our functor $S°$ is known as "the localization functor associated with the singular torsion theory." Since $S°$ is the only localization functor used in this book, we refer to it simply as "localization."

Returning to the commutative domain analogy, we still must prove that $S°$ is exact. For use in this and other situations, we develop the following concept of closure for submodules.

Definition An *$\mathscr{S}$-closed* submodule of a module C is a submodule B for which C/B is nonsingular, and we use $L^*(C)$ to denote the collection of all $\mathscr{S}$-closed submodules of C. Note that $L^*(C)$ is closed under arbitrary intersections: For if $\{B_\alpha\} \subseteq L^*(C)$, then $C/(\cap B_\alpha)$ can be embedded in the nonsingular module $\Pi(C/B_\alpha)$. Thus for any $A \leqq C$ there is a smallest $\mathscr{S}$-closed submodule of C containing A, which is called the *$\mathscr{S}$-closure of A in C*. The following proposition provides alternate descriptions of the $\mathscr{S}$-closure of A.

Proposition 2.3 Assume that $Z_r(R) = 0$. Let $A \leqq C$ be right R-modules, and let K be the $\mathscr{S}$-closure of A in C.

(a) $K/A = Z(C/A)$.

(b) K is the only $\mathscr{S}$-closed submodule of C for which $A \leqq$ K and K/A is singular.

(c) If C is nonsingular, then K is the only $\mathscr{S}$-closed submodule of C for which $A \leqq_e K$.

Proof: (a) If $T/A = Z(C/A)$, then $C/T \cong (C/A)/[Z(C/A)]$ is nonsingular by 1.23, whence $T \in L^*(C)$ and so $K \leqq T$. On the other hand, T/K is an epimorphic image of T/A and so is singular, whence $T/K \leqq Z(C/K) = 0$. Thus $T = K$.

(b) By definition, $A \leqq K \in L^*(C)$, and (a) shows that K/A is singular. Conversely, suppose that $A \leqq K' \in L^*(C)$ and K'/A is singular. Then $K \leqq K'$ by definition of K, and $K'/A \leqq Z(C/A) = K/A$ by (a), whence $K' = K$.

(c) follows from (b) by using 1.21.□

When $Z_r(R) = 0$, 2.3 shows in particular that the singular submodule of any right R-module C is just the $\mathscr{S}$-closure of 0 in C. Also, if H is any two-sided ideal of R and K is the $\mathscr{S}$-closure of H_R in R_R, then $K/H = Z((R/H)_R)$ by 2.3. Thus K/H is a fully invariant submodule of R/H, and so K is a two-sided ideal of R. Actually, this conclusion is valid even if $Z_r(R) \neq 0$ (Exercise 1).

In general, closed submodules need not be $\mathscr{S}$-closed. For example, 0 is a closed submodule of any module C, but 0 is $\mathscr{S}$-closed in C only if C is nonsingular. On the other hand, $\mathscr{S}$-closed submodules are always closed, as the next proposition shows.

Proposition 2.4 Every $\mathscr{S}$-closed submodule of a module C is closed in C. If C is nonsingular, then every closed submodule of C is $\mathscr{S}$-closed in C.

Proof: If $A \in L^*(C)$ and $A \leqq_e B \leqq C$, then $B/A \leqq Z(C/A) = 0$, whence $B = A$. Conversely, let C be nonsingular, and let A be any closed submodule of C. If $K/A = Z(C/A)$, then $A \leqq_e K$ by 1.21 and so $K = A$, whence $Z(C/A) = 0$. Thus A is $\mathscr{S}$-closed in C.□

Lemma 2.5 Let $Z_r(R) = 0$, and let $f : A \to B$ be a monomorphism of right R-modules.

(a) $\bar{f} : A/Z(A) \to B/Z(B)$ and $S^\circ f : S^\circ A \to S^\circ B$ are both monomorphisms.

(b) $(S^\circ f)(S^\circ A)$ is the $\mathscr{S}$-closure of $\bar{f}(A/Z(A))$ in $S^\circ B$.

Proof: (a) Observing that $f(Z(A)) = Z(fA) = fA \cap Z(B)$, we infer that $f^{-1}(Z(B)) = Z(A)$, whence $\bar{f}$ is a monomorphism. Then $[A/Z(A)] \cap [\ker (S^\circ f)] = 0$, from which we conclude that $\ker (S^\circ f) = 0$ also.

(b) Since $S^\circ f$ is a monomorphism by (a), $(S^\circ f)(S^\circ A)$ is isomorphic to $S^\circ A$ and hence is injective. Then $(S^\circ f)(S^\circ A)$ has no proper essential extensions, and so in particular is a closed submodule of $S^\circ B$. According to 2.4, $(S^\circ f)(S^\circ A) \in L^*(S^\circ B)$. Also, $[(S^\circ f)(S^\circ A)]/[\bar{f}(A/Z(A))]$ is isomorphic to $[S^\circ A]/[A/Z(A)]$ and thus is singular; hence by 2.3 $(S^\circ f)(S^\circ A)$ is the $\mathscr{S}$-closure of $\bar{f}(A/Z(A))$ in $S^\circ B$.□

Whenever $j : A \to B$ is an inclusion map, it is convenient, in view of 2.5, to identify $A/Z(A)$ with $\bar{j}(A/Z(A))$ and $S^\circ A$ with $(S^\circ j)(S^\circ A)$.

Theorem 2.6 If $Z_r(R) = 0$, then S° is an exact functor.

Proof: If $0 \to A \xrightarrow{f} B \xrightarrow{g} C \to 0$ is a short exact sequence of right R-modules, then by 2.5 we already know that $S^\circ f$ is a monomorphism.

Note that $g^{-1}(Z(C))/(fA) = g^{-1}(Z(C))/(\ker g)$ is isomorphic to $Z(C)$ and so is singular. Inasmuch as $g^{-1}(Z(C))/Z(B) = \ker \bar{g}$, there is an epimorphism of $g^{-1}(Z(C))/(fA)$ onto $[\ker \bar{g}]/[\bar{f}(A/Z(A))]$, whence $[\ker \bar{g}]/[\bar{f}(A/Z(A))]$ must be singular. Intersecting $\ker(S^\circ g)$ with the relationship $B/Z(B) \leqq_e S^\circ B$, we find that $\ker \bar{g} \leqq_e \ker(S^\circ g)$, whence $[\ker(S^\circ g)]/[\ker \bar{g}]$ is singular. Now 1.23 shows that $[\ker(S^\circ g)]/[\bar{f}(A/Z(A))]$ is singular. On the other hand, $S^\circ g$ induces a monomorphism of $S^\circ B/[\ker(S^\circ g)]$ into the nonsingular module $S^\circ C$; hence $\ker(S^\circ g) \in L^*(S^\circ B)$. According to 2.3, $\ker(S^\circ g)$ is thus the $\mathscr{S}$-closure of $\bar{f}(A/Z(A))$ in $S^\circ B$, whence 2.5 says that $\ker(S^\circ g) = (S^\circ f)(S^\circ A)$. Therefore $0 \to S^\circ A \to S^\circ B \to S^\circ C$ is exact.

This much exactness shows that $\ker(S^\circ g)$ is isomorphic to $S^\circ A$ and so is injective. Then $\ker(S^\circ g)$ is a direct summand of $S^\circ B$, and the complementary summand is isomorphic to $(S^\circ g)(S^\circ B)$. Thus $(S^\circ g)(S^\circ B)$ is injective too; hence it has no proper essential extensions. However, $(S^\circ g)(S^\circ B)$ contains $\bar{g}(B/Z(B)) = C/Z(C)$, which is essential in $S^\circ C$; hence $(S^\circ g)(S^\circ B) \leqq_e S^\circ C$. Therefore $(S^\circ g)(S^\circ B) = S^\circ C$.□

Inasmuch as R is a subring of $S^\circ R$, all $S^\circ R$-modules are also R-modules. In general, R-homomorphisms between $S^\circ R$-modules need not be $S^\circ R$-homomorphisms, as shown by Exercise 2. However, we are mainly concerned with right $S^\circ R$-modules A,B such that B_R is nonsingular (and usually A_R as well), and in this case the following proposition shows that there is no problem.

Proposition 2.7 Let $Z_r(R) = 0$, and let A,B be right $S^\circ R$-modules.

(a) If $Z_R(B) = 0$, then $\mathrm{Hom}_R(A,B) = \mathrm{Hom}_{S^\circ R}(A,B)$.

(b) If $Z_R(A) = 0$, then A is an $S^\circ R$-submodule of $S^\circ A$.

(c) Suppose that B has an R-submodule M such that $B = A \oplus M$. If $Z_R(A) = 0$, then M is an $S^\circ R$-submodule of B.

Proof: (a) Given $f \in \mathrm{Hom}_R(A,B)$ and $x \in A$, we may define an R-homomorphism $g : S^\circ R \to B$ by the rule $ga = f(xa) - (fx)a$. Clearly $gR = 0$; hence it follows from 2.1 that $g = 0$. Thus f is an $S^\circ R$-homomorphism.

(b) According to (a), the inclusion map $A \to S^\circ A$ is an $S^\circ R$-homomorphism.

(c) If $f : B \to A$ is the projection with kernel M, then by (a), f is an $S^\circ R$-homomorphism. Thus $M = \ker f$ is an S°R-submodule of B.□

If we are dealing with right $S^\circ R$-modules $A \leqq B$ such that $Z_R(A) = 0$, then 2.7(c) shows that the phrase "A is a direct summand of B" has the same meaning whether A and B are considered as $S^\circ R$-modules or as R-modules.

In a similar vein, we ask if the singularity or nonsingularity of a right $S^\circ R$-module A is the same whether A is considered as an $S^\circ R$-module or as an R-module. The next proposition shows that the answer is yes.

Proposition 2.8 Let $Z_r(R) = 0$, and set $Q = S^\circ R$.

(a) $\mathscr{S}(Q) = \{I \leqq Q_Q \mid I \cap R \in \mathscr{S}(R)\}$; $\mathscr{S}(R) = \{J \leqq R_R \mid JQ \in \mathscr{S}(Q)\}$.

(b) $Z_Q(A) = Z_R(A)$ for all A_Q.

(c) $Z_r(Q) = 0$.

Proof: (a) If I is a right ideal of Q such that $I \cap R \in \mathscr{S}(R)$, then since $I \cap R \leqq_e R_R \leqq_e Q_R$ we obtain $I_R \leqq_e Q_R$, whence $I_Q \leqq_e Q_Q$.

Now let $I \in \mathscr{S}(Q)$, and consider any right ideal M of R such that $M \cap (I \cap R) = 0$. Since $M \leqq_e S^\circ M$ and $I \cap R \leqq_e I$, 1.1 says that $S^\circ M \cap I = 0$; hence $S^\circ M = 0$ and so $M = 0$. Thus $I \cap R \in \mathscr{S}(R)$.

If $J \in \mathscr{S}(R)$, then since $J \leqq JQ \cap R$ we have $JQ \cap R \in \mathscr{S}(R)$; hence $JQ \in \mathscr{S}(Q)$.

Finally, let J be any right ideal of R such that $JQ \in \mathscr{S}(Q)$. Since $S^\circ J$ is a right ideal of Q, we have $JQ \leqq S^\circ J$ and thus $J \leqq_e JQ$, whence $J \leqq_e JQ \cap R$. We also have $JQ \cap R \in \mathscr{S}(R)$; hence $J \in \mathscr{S}(R)$.

(b) and (c) are clear from (a). □

Proposition 2.9 Let $Z_r(R) = 0$, and let C be any right R-module.

(a) $S^\circ C$ is a nonsingular injective right $S^\circ R$-module.

(b) Every member of $L^*(S^\circ C)$ is an $S^\circ R$-submodule of $S^\circ C$ and is a direct summand of $S^\circ C$.

(c) If A is a finitely generated $S^\circ R$-submodule of $S^\circ C$, then $A \in L^*(S^\circ C)$ and $S^\circ A = A$.

Proof: (a) In view of 2.8, $(S^\circ C)_{S^\circ R}$ is nonsingular. If B is any right $S^\circ R$-module containing $S^\circ C$, then $(S^\circ C)_R$ is a direct summand of B_R, because $(S^\circ C)_R$ is injective. Since $(S^\circ C)_R$ is nonsingular, it follows from 2.7 that $S^\circ C$ is also a direct summand of B when considered as $S^\circ R$-modules. Thus $(S^\circ C)_{S^\circ R}$ is injective.

(b) If $A \in L^*(S^\circ C)$, then $S^\circ A = A$ by 2.5, whence A is an $S^\circ R$-submodule of $S^\circ C$. According to (a), A is an injective $S^\circ R$-module; hence A is a direct summand of $S^\circ C$.

(c) Choose an epimorphism $f: F \to A$, where F is a finitely generated free right $S^\circ R$-module. Since F is isomorphic to a finite direct sum of copies of $S^\circ R$, it is a nonsingular injective right R-module, whence $S^\circ F = F$. Now $S^\circ f$ is surjective by 2.6; hence we obtain $S^\circ A = A$. According to 2.5, $A \in L^*(S^\circ C)$. □

Corollary 2.10 Let $Z_r(R) = 0$, and let C be a nonsingular right R-module. If C has a finitely generated essential submodule, then $S^\circ C = C(S^\circ R)$ and $S^\circ C$ is a finitely generated $S^\circ R$-module.

Proof: If A is a finitely generated essential submodule of C, then $A \leqq_e C \leqq_e S^\circ C$, whence $A(S^\circ R) \leqq_e S^\circ C$. On the other hand, $A(S^\circ R)$ is a finitely generated $S^\circ R$-submodule of $S^\circ C$; hence 2.9 says that $A(S^\circ R) \in L^*(S^\circ C)$. As a result, $A(S^\circ R) = S^\circ C$, from which the required conclusions are clear. □

In case $C = R$, 2.9 says that $(S^\circ R)_{S^\circ R}$ is injective and that every finitely generated right ideal of $S^\circ R$ is a direct summand of $S^\circ R$, that is, $S^\circ R$ is a right self-injective ring and a regular ring. These two properties essentially characterize $S^\circ R$, as follows.

Proposition 2.11 Assume that R is a subring of a ring Q, and that $R_R \leqq_e Q_R$. Then the following conditions are equivalent:

(a) Q is a regular, right self-injective ring.

(b) $Z_r(R) = 0$ and Q_Q is injective.

(c) $Z_r(R) = 0$ and $Q = S^\circ R$.

Proof: We have just observed that (c) ⇒ (a).

(a) ⇒ (b): According to 1.27, $Z_r(Q) = 0$. As in 2.8, we have $IQ \in \mathscr{S}(Q)$ for all $I \in \mathscr{S}(R)$, whence $Z_r(R) \subseteq Z_r(Q) = 0$.

(b) ⇒ (c): According to 1.7, the inclusion map $R \to S^\circ R$ extends to a monomorphism $f : Q_R \to (S^\circ R)_R$, and we claim that f is actually a ring morphism. Given any $x \in Q$, we define an R-homomorphism $g : Q_R \to (S^\circ R)_R$ by the rule $gy = f(xy) - (fx)(fy)$. Observing that $gR = 0$, we obtain $g = 0$ from 2.1. Therefore f is an isomorphism of Q onto the subring fQ of $S^\circ R$; hence we may identify Q with fQ. Since $R_R \leqq_e (S^\circ R)_R$, we have $Q_R \leqq_e (S^\circ R)_R$ and thus $Q_Q \leqq_e (S^\circ R)_Q$. Inasmuch as Q_Q is injective, we conclude from 1.8 that $Q = S^\circ R$. □

Proposition 2.11 is particularly useful for identifying $S^\circ R$ when constructing examples (e.g., Exercises 3 and 4). This requires a large source of regular, right self-injective rings, which we develop in the next section. We show, for example, that the endomorphism ring of any nonsingular quasi-injective right module is regular and right self-injective.

We conclude the present section by showing that homomorphisms into $S^\circ R$ determine the singularity or nonsingularity of all right R-modules.

Proposition 2.12 Assume that $Z_r(R) = 0$, and let A be a right R-module.

(a) If A is nonsingular and has a finitely generated essential submodule, then A can be embedded in a finite direct sum of copies of $S^\circ R$.

(b) A is singular if and only if $\mathrm{Hom}_R(A,S^\circ R) = 0$.

(c) A is nonsingular if and only if A can be embedded in a direct product of copies of $S^\circ R$.

Proof: (a) According to 2.10, $S^\circ A$ is a finitely generated right $S^\circ R$-module; hence we can choose an epimorphism $f : F \to S^\circ A$, where F is a finite direct sum of copies of $S^\circ R$. Since $S^\circ A$ is nonsingular, $\ker f \in L^*(F)$; hence by 2.4 $\ker f$ is closed in F. However, F is injective because $S^\circ R$ is; hence 1.9 says that $\ker f$ is injective. Thus $S^\circ A$ is isomorphic to a direct summand of F, which gives us an embedding $A \to S^\circ A \to F$.

(b) If A is singular, then $\mathrm{Hom}_R(A,S^\circ R) = 0$ by 1.20. If A is not singular, then $A/Z(A)$ has a nonzero element x. Now $\mathrm{Hom}_R(xR,S^\circ R) \neq 0$ by (a); hence it follows from the injectivity of $S^\circ R$ that $\mathrm{Hom}_R(A,S^\circ R) \neq 0$.

(c) If A can be embedded in a direct product of copies of $S^\circ R$, then A is nonsingular by 1.22. Conversely, assume that A is nonsingular, and let K denote the intersection of the kernels of all maps from $A \to S^\circ R$. Noting that the natural map $\mathrm{Hom}_R(A,S^\circ R) \to \mathrm{Hom}_R(K,S^\circ R)$ is zero, we infer from the injectivity of $S^\circ R$ that $\mathrm{Hom}_R(K,S^\circ R) = 0$. Then K is singular by (b) and so $K = 0$, which yields the required embedding. □

Exercises

1. If H is a two-sided ideal of R and K is the $\mathscr{S}$-closure of H_R in R_R, prove that K is a two-sided ideal.
2. Let F be a field, $F_n = F$ for $n = 1,2,\ldots$, $Q = \Pi F_n$, $J = \oplus F_n$. If R is the F-subalgebra of Q generated by 1 and J, show that $Z_r(R) = 0$ and $S^\circ R = Q$. Also, show that Q/J is a singular Q-module and that $\mathrm{End}_Q(Q/J) \neq \mathrm{End}_R(Q/J)$.
3. If R is a commutative integral domain, show that $S^\circ R$ is the quotient field of R, and that S° is naturally equivalent to the functor S^{-1} described at the beginning of the section.
4. Let Q be a simple artinian ring. If R is a subring of Q which contains a nonzero *left* ideal of Q, prove that $Z_r(R) = 0$ and $S^\circ R = Q$.
5. Let $A \leqq B \leqq C$ and assume that C is nonsingular. Show that $A \leqq_e B$ if and only if B is contained in the $\mathscr{S}$-closure of A in C.
6. Let R be commutative and nonsingular. Show that the $\mathscr{S}$-closure of any ideal I in R is the annihilator of the annihilator of I.
7. If $f : B \to C$ and $A \in L^*(C)$, show that $f^{-1}A \in L^*(B)$.
8. If $A \leqq B \leqq C$ and $A \in L^*(B)$, $B \in L^*(C)$, show that $A \in L^*(C)$.
9. If $A_\alpha \in L^*(B_\alpha)$ for each α in some index set, show that $\oplus A_\alpha \in L^*(\oplus B_\alpha)$ and $\Pi A_\alpha \in L^*(\Pi B_\alpha)$.
10. Find an example of a module C such that $L^*(C)$ is not closed under finite sums.
11. For any module A, prove that $L^*(A)$ is a complete, complemented, modular lattice.
12. Suppose that $R = \Pi R_\alpha$ and that $Z_r(R_\alpha) = 0$ for all α, so that $Z_r(R) = 0$ (Exercise 1.D.12). Show that $S^\circ R = \Pi(S^\circ R_\alpha)$.

13. Considering $S°$ as a functor from Mod-R to Mod-R, show that $S°S° = S°$.
14. Let $Z_r(R) = 0$. Prove that an element of $S°R$ is central if and only if it commutes with all elements of R. Conclude that $S°R$ is commutative if and only if R is commutative.
15. If $Z_r(R) = 0$, prove that any ring automorphism of R extends uniquely to a ring automorphism of $S°R$.
16. If $Z_r(R) = 0$, prove that any derivation of R extends uniquely to a derivation of $S°R$.
17. If $Z_r(R) = 0$ and ${}_RR \leqq_e {}_R(S°R)$, prove that any involution of R extends uniquely to an involution of $S°R$.
18. If $Z_r(R) = 0$, show that $L^*(R_R) = \{R \cap e(S°R) \mid e = e^2 \in S°R\}$.
19. Let $Z_r(R) = 0$, and let A be any right R-module. Show that the kernel of the natural map $A \to A \otimes_R S°R$ is contained in $Z(A)$. Thus if $A \otimes_R S°R = 0$, A must be singular. With notation as in Exercise 2, set $A = R/J$ and show that A is singular but that $A \otimes_R S°R \neq 0$.
20. If $Z_r(R) = 0$ and A is a countably generated nonsingular right R-module, prove that A can be embedded in a countable direct sum of copies of $S°R$.
21. For any right R-module A, show that $Z_2(A)$ is the $\mathscr{S}$-closure of 0 in A. (See Exercise 1.D.20.)
22. For any ring R, we know that $J = Z_2(R_R)$ is a two-sided ideal of R (Exercise 1.D.20). Prove that every nonsingular right R-module is also a nonsingular right (R/J)-module, and vice versa. Prove that a nonsingular right R-module is injective as an R-module if and only if it is injective as an (R/J)-module.
23. Prove that all singular right R-modules are injective if and only if $Z_r(R) = 0$ and R/I is semisimple for all $I \in \mathscr{S}(R)$. Show that the ring R in Exercise 2 is an example of a nonsemisimple ring satisfying these conditions.
24. If all singular right R-modules are injective, prove that there exists an additive functor F : Mod-$R \to$ Mod-R such that $F(A) \cong E(A)$ for all A_R.

B. Endomorphism Rings of Quasi-injective Modules

Definition A module A is said to be *quasi-injective* provided the natural map $\mathrm{Hom}_R(A,A) \to \mathrm{Hom}_R(M,A)$ is surjective for all $M \leqq A$, i.e., provided any homomorphism from a submodule of A into A extends to an endomorphism of A. Obviously all injective modules are quasi-injective. Also, any semisimple module A is quasi-injective: for any $M \leqq A$, the sequence $0 \to M \to A \to A/M \to 0$ is split exact, whence $\mathrm{Hom}_R(A,A) \to \mathrm{Hom}_R(M,A)$ is surjective. Thus, for example, $\mathbf{Z}/2\mathbf{Z}$ is an example of a quasi-injective $\mathbf{Z}$-module which is not injective.

Proposition 2.13 A right R-module A is quasi-injective if and only if A is a fully invariant submodule of $E(A)$.

Proof: Set $T = \mathrm{End}_R(E(A))$.

First, assume that $TA \leqq A$. Given $M \leqq A$, any $f: M \to A$ must extend to some $t \in T$, whence $t|_A$ is an endomorphism of A which extends f. Thus A is quasi-injective.

Now let A be quasi-injective, and let $t \in T$. Restricting t, we get a map from $A \cap t^{-1}A$ into A, which by quasi-injectivity extends to an endomorphism f of A. Then f extends to a map $g \in T$ such that $gA \leqq A$ and $(g - t)(A \cap t^{-1}A) = 0$. Since $gA \leqq A$, we infer that

$$A \cap (g - t)^{-1}A \leqq A \cap t^{-1}A \leqq \ker(g - t)$$

whence $(g - t)A \cap A = 0$. Then $(g - t)A = 0$ [because $A \leqq_e E(A)$]; hence we conclude that $tA = gA \leqq A$. Therefore $TA \leqq A$.□

Corollary 2.14 If A is a quasi-injective module, then any decomposition $E(A) = \oplus E_\alpha$ induces a corresponding decomposition $A = \oplus(A \cap E_\alpha)$.

Proof: For each α, let $p_\alpha : E(A) \to E(A)$ be the projection onto the direct summand E_α. Since $p_\alpha A \leqq A$ by 2.13, we see that the E_α-component of any element of A also belongs to A, from which we obtain $A = \oplus(A \cap E_\alpha)$.□

Corollary 2.15 Let C be a quasi-injective module. Then all closed submodules of C are direct summands of C, and all direct summands of C are quasi-injective.

Proof: Let A be a closed submodule of C, and choose injective hulls $E(A) \leqq E(C)$. Inasmuch as $A \leqq_e C \cap E(A) \leqq C$, we obtain $C \cap E(A) = A$. Now $E(C) = E(A) \oplus B$ for some B; hence 2.14 says that

$$C = [C \cap E(A)] \oplus (C \cap B) = A \oplus (C \cap B)$$

Next, consider any direct sum decomposition $C = A \oplus B$. Choose a corresponding decomposition $E(C) = E(A) \oplus E(B)$, and let $T = \text{End}_R(E(C))$. If $p \in T$ is the projection onto $E(A)$, then $pTp = \text{End}_R(E(A))$. Now $TC \leqq C$ by 2.13, whence $pTpC \leqq pC$ and so $pTpA \leqq A$. Using 2.13 again, we find that A is quasi-injective.□

Unlike the situation for injective modules, direct products of quasi-injective modules need not be quasi-injective. For example, $\mathbf{Z}/2\mathbf{Z}$ and $\mathbf{Q}$ are quasi-injective $\mathbf{Z}$-modules, but $(\mathbf{Z}/2\mathbf{Z}) \oplus \mathbf{Q}$ is not quasi-injective, since the map $(0, x) \mapsto (\bar{x}, 0)$ from $0 \oplus \mathbf{Z}$ to $(\mathbf{Z}/2\mathbf{Z}) \oplus \mathbf{Q}$ does not extend to an endomorphism of $(\mathbf{Z}/2\mathbf{Z}) \oplus \mathbf{Q}$. (See also Exercise 1.) In general, the most that can be said is that for any quasi-injective module A, finite direct products of copies of A are quasi-injective (Exercise 2).

Theorem 2.16 Let A be a quasi-injective right R-module, and set $Q = \text{End}_R(A)$. Then $J(Q) = \{f \in Q \mid \ker f \leqq_e A\}$, and $Q/J(Q)$ is a regular ring.

Proof: Set $K = \{f \in Q \mid \ker f \leqq_e A\}$, and consider any $f,g \in K$. Since $(\ker f) \cap (\ker g) \leqq \ker(f - g)$, we infer that $\ker(f - g) \leqq_e A$, whence $f - g \in K$. Given any $h \in Q$, we have $\ker(fh) = h^{-1}(\ker f) \leqq_e A$, whence $fh \in K$. Also, since $\ker f \leqq \ker(hf)$, we see that $hf \in K$ as well. Therefore K is a two-sided ideal of Q.

Given any $f \in K$, we have $\ker f \leqq_e A$ and $[\ker(1 - f)] \cap [\ker f] = 0$; hence $\ker(1 - f) = 0$. Then $1 - f$ provides an isomorphism of A onto $(1 - f)A$, and the inverse isomorphism $(1 - f)A \to A$ extends to a map $g \in Q$ such that $g(1 - f) = 1$. Thus f is a left quasi-regular element of Q. Now K is a left quasi-regular ideal of Q, and so $K \leqq J(Q)$.

Before showing that $K = J(Q)$, we first prove that Q/K is a regular ring, from which $K = J(Q)$ follows easily.

Thus consider any $f \in Q$, and let B be a relative complement for $\ker f$ in A. Noting that f restricts to an isomorphism of B onto fB, we use quasi-injectivity again to extend the inverse isomorphism $fB \to B$ to some $g \in Q$. Now $(gf)|_B$ is the identity on B; hence $(fgf - f)B = 0$, and consequently $B \oplus (\ker f) \leqq \ker(fgf - f)$. Inasmuch as $B \oplus (\ker f) \leqq_e A$, we thus obtain $fgf - f \in K$, whence $\overline{fgf} = \bar{f}$ in Q/K. Therefore Q/K is a regular ring.

Now regular rings have zero radical, hence $J(Q/K) = 0$. On the other hand, since $K \leqq J(Q)$ we have $J(Q/K) = J(Q)/K$, whence $J(Q) = K$. Consequently, $Q/J(Q)$ is regular. □

In the situation described in 2.16, it is also true that $Q/J(Q)$ is a right self-injective ring. However, some preparation is needed before proving this result. First, the following corollary shows that we can reduce to the case where A is injective. Second, we must prove some results about "lifting" idempotents from $Q/J(Q)$ to idempotents in Q.

Corollary 2.17 Let A be a quasi-injective right R-module, and set $Q = \mathrm{End}_R(A)$, $T = \mathrm{End}_R(E(A))$. Then $Q/J(Q) \cong T/J(T)$.

Proof: In view of 2.13, we have a ring morphism $\phi : T \to Q$ given by the rule $\phi(f) = f|_A$. Since $E(A)$ is injective, we see that ϕ is surjective, whence $T/[\phi^{-1}(J(Q))] \cong Q/J(Q)$. Thus it suffices to prove that $\phi^{-1}(J(Q)) = J(T)$.

Given any $f \in T$, we see from 2.16 that $f \in \phi^{-1}(J(Q))$ if and only if $\ker(f|_A) \leqq_e A$, that is, if and only if $A \cap (\ker f) \leqq_e A$. This happens if and only if $\ker f \leqq_e E(A)$, which by 2.16 is equivalent to having $f \in J(T)$. Therefore $\phi^{-1}(J(Q)) = J(T)$, and we are done. □

Given a quasi-injective module A, it is in general not true that $\mathrm{End}_R(A) \cong \mathrm{End}_R(E(A))$ (Exercise 3).

Lemma 2.18 Let A be an injective module, and set $Q = \mathrm{End}_R(A)$. If x is any idempotent in $\bar{Q} = Q/J(Q)$, then there exists an idempotent $e \in Q$ such that $\bar{e} = x$.

Proof: First, $x = \bar{f}$ for some $f \in Q$, and then $f - f^2 \in J(Q)$; hence 2.16 says that $\ker(f - f^2) \leqq_e A$. Observing that $f[\ker(f - f^2)] \leqq \ker(1 - f)$ and that $(1 - f)[\ker(f - f^2)] \leqq \ker f$, we infer that

$$\ker(f - f^2) \leqq [\ker(1 - f)] \oplus [\ker f]$$

whence $[\ker(1 - f)] \oplus [\ker f] \leqq_e A$. Since A is injective, we obtain from this a decomposition $A = A_1 \oplus A_2$ such that $A_1 = E(\ker(1 - f))$ and $A_2 = E(\ker f)$. If $e \in Q$ denotes the projection onto A_1, then $e = e^2$ and we observe that e, f agree on $[\ker(1 - f)] \oplus [\ker f]$. Thus $\ker(e - f) \leqq_e A$; hence by 2.16 we have $e - f \in J(Q)$, and finally $\bar{e} = \bar{f} = x$.□

Lemma 2.19 Let A be an injective module, and set $Q = \mathrm{End}_R(A)$. If e,f are idempotents in Q such that $\overline{eQ} \cap \overline{fQ} = 0$ in $Q/J(Q)$, then $eA \cap fA = 0$.

Proof: We first must change f slightly in order to obtain $\overline{fe} = 0$. Since $\bar{Q}$ is a regular ring by 2.16, we must have $\bar{Q}_{\bar{Q}} = \overline{eQ} \oplus \overline{fQ} \oplus I$ for some I. There is an idempotent $x \in \bar{Q}$ such that $x\bar{Q} = \overline{fQ}$ and $(1 - x)\bar{Q} = \overline{eQ} \oplus I$, and by 2.18 there is an idempotent $h \in Q$ with $\bar{h} = x$. Note that $\overline{hQ} = \overline{fQ}$ and that $\overline{he} = 0$.

Since $\overline{hQ} = \overline{fQ}$, we have $\bar{h} = \overline{fh}$ and $\bar{f} = \overline{hf}$, from which we compute that $\bar{f} + \overline{fh}(1 - \bar{f}) = \bar{h}$. Setting $g = f + fh(1 - f)$, we thus have $\bar{g} = \bar{h}$; and since $f = f^2$ we also compute that $g = g^2$. Noting that $fg = g$ and $gf = f$, we see that $fA = gA$.

We now have an idempotent $g \in Q$ such that $\overline{gQ} = \overline{hQ} = \overline{fQ}$, $\overline{ge} = \overline{he} = 0$, and $gA = fA$. Then $\overline{eQ} \cap \overline{gQ} = 0$, and it suffices to prove that $eA \cap gA = 0$. Therefore, replacing f by g, we may assume (without loss of generality) that $\overline{fe} = 0$.

Now $fe \in J(Q)$ and thus $\ker(fe) \leqq_e A$, by 2.16. Setting $K = \ker(fe)$, we infer that $K \cap eA \leqq_e eA$, from which it follows that $eK \leqq_e eA$. Inasmuch as $feK = 0$, we see that $eK \cap fA = 0$, from which we conclude that $eA \cap fA = 0$.□

Lemma 2.20 Let A be an injective module, and set $Q = \mathrm{End}_R(A)$. If $\{e_\alpha\}$ is a collection of idempotents in Q such that $\{\bar{e}_\alpha \bar{Q}\}$ is an independent family of right ideals of $Q/J(Q)$, then $\{e_\alpha A\}$ is an independent family of submodules of A.

Proof: It suffices to check the case when the index set is finite, say $\{1, 2, \ldots, n\}$. If $n = 1$, there is nothing to prove; while if $n = 2$, we just quote 2.19.

Now let $n > 2$, and assume that $\{e_1 A, \ldots, e_{n-1} A\}$ is independent. Each

e_iA is a direct summand of A and so is injective; hence $e_1A \oplus \cdots \oplus e_{n-1}A$ is injective. Then $e_1A \oplus \cdots \oplus e_{n-1}A = eA$ for some idempotent $e \in Q$, and we claim that $eQ = e_1Q + \cdots + e_{n-1}Q$. Since $ee_i = e_i$ for $i = 1, \ldots, n-1$, we have $e_1Q + \cdots + e_{n-1}Q \leqq eQ$. For each $i = 1, \ldots, n-1$, let f_i denote the composition of e with the projection $eA \to e_iA$. Then $e_if_i = f_i$ for each i, and $f_1 + \cdots + f_{n-1} = e$, from which we obtain $eQ = e_1Q + \cdots + e_{n-1}Q$, as claimed.

Passing to $Q/J(Q)$, we get $\overline{eQ} = \bar{e}_1\bar{Q} + \cdots + \bar{e}_{n-1}\bar{Q}$, whence $\overline{eQ} \cap \bar{e}_n\bar{Q} = 0$. According to 2.19, $eA \cap e_nA = 0$, from which we conclude that $\{e_1A, \ldots, e_nA\}$ is independent.□

Theorem 2.21 If A is a quasi-injective right R-module and $Q = \mathrm{End}_R(A)$, then $Q/J(Q)$ is a regular, right self-injective ring.

Proof: In view of 2.17, we need only consider the case when A is actually injective. Also, $Q/J(Q)$ is already regular, by 2.16.

We must show that for any nonzero right ideal I of $\bar{Q}$, any right $\bar{Q}$-homomorphism $f : I \to \bar{Q}$ is given by left multiplication by some element of $\bar{Q}$. First choose a maximal independent family $\{C_\alpha\}$ of nonzero cyclic submodules of I, and note that $\oplus C_\alpha \leqq_e I$. Since $\bar{Q}$ is regular, each C_α can be generated by an idempotent, and according to 2.18 these idempotents all lift to idempotents in Q. Thus we obtain a collection $\{e_\alpha\}$ of idempotents in Q such that $\{\bar{e}_\alpha\bar{Q}\}$ is an independent family of right ideals of $\bar{Q}$, and $\oplus\bar{e}_\alpha\bar{Q} \leqq_e I$.

According to 2.20, $\{e_\alpha A\}$ is an independent family of submodules of A. For each α, choose $t_\alpha \in Q$ such that $\bar{t}_\alpha = f(\bar{e}_\alpha)$, and restrict t_α to a map $e_\alpha A \to A$. Together, these maps induce a map $\oplus e_\alpha A \to A$, and by injectivity this extends to a map $w \in Q$. Inasmuch as w agrees with t_α on $e_\alpha A$, we obtain $we_\alpha = t_\alpha e_\alpha$ for each α.

Now $\overline{we_\alpha} = (f\bar{e}_\alpha)\bar{e}_\alpha = f(\bar{e}_\alpha)$ for each α; hence left multiplication by $\bar{w}$ agrees with f on $\oplus\bar{e}_\alpha\bar{Q}$. Inasmuch as $Z_r(\bar{Q}) = 0$ by 1.27, we conclude using 2.1 that f is given by left multiplication by $\bar{w}$.□

Corollary 2.22 Let A be a right R-module, and set $Q = \mathrm{End}_R(A)$. If A is either semisimple or nonsingular quasi-injective, then Q is a regular, right self-injective ring.

Proof: Since A is quasi-injective in either case, it follows from 2.21 that we need only show $J(Q) = 0$. Given any $f \in J(Q)$, we have $\ker f \leqq_e A$ by 2.16. If A is semisimple, then $\ker f = A$ by 1.15 and so $f = 0$. On the other hand, if A is nonsingular, we obtain $f = 0$ from 2.1. Therefore $J(Q) = 0$ in either case.□

Proposition 2.23 Let V be a right vector space over a division ring D. If

$R = \text{End}_D(V)$, then R is a regular, right self-injective ring. However, R is left self-injective if and only if V is finite-dimensional.

Proof: Since V is a semisimple right D-module, 2.22 shows that R is regular and right self-injective. If $[V : D] < \infty$, then R is semisimple and hence is left self-injective as well.

Assuming conversely that $[V : D] = \infty$, we can find an infinite linearly independent sequence $\{v_1, v_2, \ldots\}$ in V. Write $V = V_0 \oplus (\oplus v_i D)$ for some V_0, and define idempotents $e_0, e_1, \ldots$ in R as follows: e_0 is the projection on V_0, while for $i > 0$, e_i is the projection on $v_i D$. Since the e_i are orthogonal idempotents with $\oplus e_i V = V$, we see that the left ideal $J = \oplus Re_i$ has zero right annihilator in R.

Now define $p \in R$ by setting $pV_0 = 0$ and $pv_i = v_1$ for all $i > 0$. We claim that $J \cap Rp = 0$. Given any $f \in R$ such that $fp \in J$, we must have $fp = r_0 e_0 + \cdots + r_n e_n$ for some $n \geqq 0$ and some $r_i \in R$. For $i = 1, \ldots, n$, we observe that

$$r_i e_i v_i = fpv_i = fv_1 = fpv_{n+1} = 0$$

whence $r_i e_i = 0$. Then $fp = r_0 e_0$, and as a result, $fp = fpe_0 = 0$. Thus $J \cap Rp = 0$.

Since $J \cap Rp = 0$, we see that $J \nleqq_e {}_R R$. If K denotes the $\mathscr{S}$-closure of J in ${}_R R$, then $J \leqq_e K$ by 2.3, whence $K \neq R$. Thus (using 2.4), K is a proper closed left ideal of R. Since the right annihilator of K is contained in the right annihilator of J and so is zero, we conclude that K cannot be a direct summand of ${}_R R$. Therefore K is a closed submodule of ${}_R R$ which is not injective; hence 1.9 shows that ${}_R R$ is not injective. □

Exercises

1. Let A, B be quasi-injective modules such that $E(A) \cong E(B)$. Prove that $A \oplus B$ is quasi-injective if and only if $A \cong B$.
2. If A is a quasi-injective module, prove that A^n is quasi-injective for all $n > 0$.
3. Find an example of a quasi-injective module A such that $\text{End}_R(A)$ is not isomorphic to any subring of $\text{End}_R(E(A))$.
4. Let A be a right R-module which has a submodule isomorphic to R_R. Prove that A is quasi-injective if and only if it is injective.
5. Let A be a nonsingular semisimple right R-module, and let B be a quasi-injective right R-module. If B is singular, or if $\text{soc}(B) = 0$, prove that $A \oplus B$ is quasi-injective.
6. Prove that all right R-modules are quasi-injective if and only if R is a semisimple ring.
7. If p is a prime number and n is a positive integer, show that $\mathbf{Z}/p^n\mathbf{Z}$ is a quasi-injective $\mathbf{Z}$-module.

8. If R is a commutative integral domain, show that all torsion-free quasi-injective R-modules are injective.
9. A *quasi-injective hull* for a module A is a quasi-injective module C, containing A, such that any monomorphism from A into a quasi-injective module K extends to a monomorphism $C \to K$. Prove that A has a quasi-injective hull. If C,C' are both quasi-injective hulls for A, prove that the identity map on A extends to an isomorphism of C onto C'.
10. If C is a quasi-injective module and A,B are closed submodules of C such that $A \cap B = 0$, prove that $A \oplus B$ is closed in C.
11. If A is any nonsingular quasi-injective module, prove that $L^*(A)$ is closed under finite sums.
12. If A is any quasi-injective module, prove that $A/Z_2(A)$ is quasi-injective. (See Exercise 1.D.20.)
13. If A is a nonsingular quasi-injective module, show that $\mathrm{End}_R(A) \cong \mathrm{End}_R(E(A))$.
14. Let A be a quasi-injective module, and set $Q = \mathrm{End}_R(A)$. If $x_1,x_2,\ldots$ is a countable sequence of orthogonal idempotents in $Q/J(Q)$, prove that there exist pairwise orthogonal idempotents $e_1,e_2,\ldots \in Q$ such that $\bar{e}_n = x_n$ for all n.
15. If R is a right self-injective ring, show that $Z_r(R) = J(R)$, and that $R/J(R)$ is a regular, right self-injective ring.
16. In case A is a nonsingular injective module, use the Hom-$\otimes$ adjoint isomorphism to prove that $\mathrm{End}_R(A)$ is right self-injective, without using idempotents as in 2.21.
17. Let V be a right vector space over a division ring D, and set $Q = \mathrm{End}_D(V)$. If R is a subring of Q which contains a nonzero *left* ideal of Q, show that $Z_r(R) = 0$ and $S^\circ R = Q$.
18. Let H be a Hilbert space, and let R be the ring of all bounded linear operators on H. Show that $Z_r(R) = Z_l(R) = 0$, and find $S^\circ R$.
19. Let R be a right noetherian, right nonsingular ring, and let F be a free right R-module. If $T = \mathrm{End}_R(F)$, prove that $Z_r(T) = 0$ and that $S^\circ T$ is isomorphic to the ring of endomorphisms of a free right $S^\circ R$-module.
20. Let R be right artinian. If A is a faithful quasi-injective right R-module, prove that A is injective.
21. Find an example of a faithful quasi-injective module which is not injective.

C. Maximal Quotient Rings

Localization in commutative ring theory (as described at the beginning of Section A) involves a formal construction of fractions or quotients; hence the rings obtained are often referred to as "rings of quotients." For example, formally inverting various multiplicatively closed subsets of a commutative integral domain R yields various subrings of the quotient field K, so that K can be thought of as a "maximal" ring of quotients of R. For purposes of generalization, we ignore the inversion of nonzero elements of R to form K, and concentrate on one consequence—that $R_R \leq_e K_R$. Thus in general one might study overrings of a ring R which are essential extensions of R (as R-modules). This procedure is fine for nonsingular rings, but for other rings we need a type of extension slightly stronger than essential extensions, as follows.

Definition Let $A \leqq B$. We say that B is a *rational extension* of A provided $\mathrm{Hom}_R(M/A,B) = 0$ whenever $A \leqq M \leqq B$, and we write $A \leqq_r B$ to denote this property. In this case, A is also called a *dense submodule* (or a *rational submodule*) of B. Note that we always have $A \leqq_r A$. As an example, consider again the $\mathbf{Z}$-submodules of $\mathbf{Q}$. If $A \leqq B \leqq \mathbf{Q}$ and $A \neq 0$, then B/A is torsion but B is torsion-free, from which we see that $A \leqq_r B$.

Lemma 2.24 (a) If $A \leqq_r B$, then $A \leqq_e B$.

(b) If B is nonsingular and $A \leqq_e B$, then $A \leqq_r B$.

Proof: (a) If $M \leqq B$ with $M \cap A = 0$, then there is an isomorphism $f:(M \oplus A)/A \to M$. Since $A \leqq_r B$, we have $\mathrm{Hom}_R((M \oplus A)/A,B) = 0$, whence $f = 0$ and so $M = 0$. Thus $A \leqq_e B$.

(b) If $A \leqq M \leqq B$, then M/A is singular (because $A \leqq_e M$), and so $\mathrm{Hom}_R(M/A,B) = 0$, by 1.20. Therefore $A \leqq_r B$.□

In general, essential extensions need not be rational. For example, set $B = \mathbf{Z}/4\mathbf{Z}$, $A = 2\mathbf{Z}/4\mathbf{Z}$, and note that $A \leqq_e B$. However, $B/A \cong A$ and so $\mathrm{Hom}_{\mathbf{Z}}(B/A,B) \neq 0$, whence $A \nleqq_r B$.

Proposition 2.25 (a) If $A \leqq B \leqq C$, then $A \leqq_r C$ if and only if $A \leqq_r B \leqq_r C$.

(b) If $A \leqq_r B \leqq C$ and $A' \leqq_r B' \leqq C$, then $A \cap A' \leqq_r B \cap B'$.

(c) For right R-modules $A \leqq B$, we have $A \leqq_r B$ if and only if the following condition holds: Whenever $x,y \in B$ with $x \neq 0$, there exists an element $r \in R$ such that $xr \neq 0$ and $yr \in A$.

Proof: (a) If $A \leqq_r C$, then it is clear from the definition that $A \leqq_r B$. Given $B \leqq M \leqq C$, $\mathrm{Hom}_R(M/B,C)$ is isomorphic to a subgroup of $\mathrm{Hom}_R(M/A,C)$, whence $\mathrm{Hom}_R(M/B,C) = 0$. Thus $B \leqq_r C$.

Conversely, assume that $A \leqq_r B \leqq_r C$, and let $A \leqq M \leqq C$. It suffices to show that any map $f: M \to C$ satisfying $fA = 0$ must be zero. Now $A \leqq M \cap B \cap f^{-1}B \leqq B$, and f restricts to a map $f': M \cap B \cap f^{-1}B \to B$ such that $f'A = 0$, hence $A \leqq_r B$ implies that $f' = 0$. Consequently, $B \cap f(M \cap B) = f'(M \cap B \cap f^{-1}B) = 0$. Since $B \leqq_r C$ implies $B \leqq_e C$, it follows that $f(M \cap B) = 0$. Now f induces a map

$$f'' : (M + B)/B \cong M/(M \cap B) \to C$$

and because $B \leqq_r C$ we obtain $f'' = 0$, whence $f = 0$. Therefore $A \leqq_r C$.

(b) Let $A \cap A' \leqq M \leqq B \cap B'$, and consider any map $f: M \to B \cap B'$ for which $f(A \cap A') = 0$. Inasmuch as $(M \cap A) \cap A' = A \cap A'$, f induces a map

$$f': [(M \cap A) + A']/A' \cong (M \cap A)/(A \cap A') \to B \cap B' \leqq B'$$

Since $A' \leqq (M \cap A) + A' \leqq B'$ and $A' \leqq_r B'$, we obtain $f' = 0$, whence $f(M \cap A) = 0$. Consequently, f induces another map

$$f'': (M + A)/A \cong M/(M \cap A) \to B \cap B' \leqq B$$

Since $A \leqq M + A \leqq B$ and $A \leqq_r B$, we must have $f'' = 0$ and so $f = 0$. Therefore $A \cap A' \leqq_r B \cap B'$.

(c) Given $A \leqq_r B$ and $x,y \in B$ with $x \neq 0$, set $J = \{r \in R \mid yr \in A\}$. Since $R/J \cong (A + yR)/A$, we must have $\mathrm{Hom}_R(R/J,B) = 0$, whence $\{b \in B \mid bJ = 0\} = 0$. Then $xJ \neq 0$; hence for a suitable $r \in J$ we obtain $xr \neq 0$ and $yr \in A$.

Conversely, if $A \nleqq_r B$ there must be a module M such that $A \leqq M \leqq B$ and $\mathrm{Hom}_R(M/A,B) \neq 0$. Choosing a nonzero map $f: M/A \to B$ and an element $y \in M$ such that $x = f\bar{y}$ is nonzero, we see that $xJ = 0$, where $J = \{r \in R \mid yr \in A\}$. Thus there are no elements $r \in R$ for which $xr \neq 0$ and $yr \in A$. □

Proposition 2.25 is analogous to 1.1, but less complete, for direct sums and inverse images of rational extensions need not be rational, as shown by Exercises 1 and 2.

Definition In general, the phrase "rational extension of A" refers to any monomorphism $f: A \to B$ such that $fA \leqq_r B$. In this case, we say that f is a *dense monomorphism*. Such an extension is called a *proper rational extension of A* provided fA is a proper submodule of B.

Analogous to the injective hull, we also have a maximal *rational* extension of any module, as in the following theorem.

Theorem 2.26 Given a module A, set $T = \mathrm{End}_R(E(A))$, and let $E_r(A) = \cap\{\ker f \mid f \in T \text{ and } fA = 0\}$.

(a) $A \leqq_r E_r(A)$, and all submodules of $E(A)$ which are rational extensions of A are contained in $E_r(A)$.

(b) Whenever $A \leqq_r B$, the inclusion map $A \to E_r(A)$ extends to a monomorphism $B \to E_r(A)$.

(c) $E_r(A)$ has no proper rational extensions.

Proof: (a) If $A \leqq M \leqq E_r(A)$ and $f: M \to E_r(A)$ with $fA = 0$, then f extends to a map $g \in T$ such that $gA = 0$. By definition, $E_r(A) \leqq \ker g$, whence $fM = gM = 0$. Thus $A \leqq_r E_r(A)$.

Now let $A \leqq_r B \leqq E(A)$, and consider any $f \in T$ for which $fA = 0$. Inasmuch as $A \leqq B \cap f^{-1}B \leqq B$ and f restricts to a map $f' : B \cap f^{-1}B \to B$ with $f'A = 0$, we must have $f' = 0$. Thus $B \cap fB = f'(B \cap f^{-1}B) = 0$ and so $fB = 0$ (because $B \leqq_e E(A)$). Therefore $B \leqq E_r(A)$.

(b) Inasmuch as $A \leqq_e B$ by 2.24, the inclusion map $A \to E(A)$ extends to a monomorphism $f : B \to E(A)$. Observing that $A = fA \leqq_r fB$, we conclude from (a) that $fB \leqq E_r(A)$.

(c) Given any dense monomorphism $f : E_r(A) \to C$, we note that $fA \leqq_r f(E_r(A)) \leqq_r C$, whence $fA \leqq_r C$. As a result, $fA \leqq_e C$; hence the isomorphism $fA \to A$ extends to a monomorphism $g : C \to E(A)$, by 1.7. Since $A = gfA \leqq_r gC$, we obtain $gC \leqq E_r(A)$ from (a). Thus we can define the composition $fg : C \to C$, and we note that $(fg - 1)(fA) = 0$. Inasmuch as $fA \leqq_r C$, we obtain $fg - 1 = 0$; hence $C = fgC \leqq f(E_r(A))$. Therefore f is not a proper rational extension of $E_r(A)$.□

Definition A *right quotient ring of* R is a ring Q which contains R as a subring such that $R_R \leqq_r Q_R$. In particular, R is always a right quotient ring of itself. As we noted above, $\mathbf{Z_Z} \leqq_r \mathbf{Q_Z}$; hence $\mathbf{Q}$ is a right quotient ring of $\mathbf{Z}$. *Left quotient rings* are defined similarly, but in general right quotient rings need not also be left quotient rings. (See Exercise 3.)

Proposition 2.27 Let R be a subring of a ring Q. If $Z_r(R) = 0$, then Q is a right quotient ring of R if and only if $R_R \leqq_e Q_R$.

Proof: This is immediate from 2.24.□

For example, if $Z_r(R) = 0$ then R is a subring of $S^\circ R$ and $R_R \leqq_e (S^\circ R)_R$; hence $S^\circ R$ is a right quotient ring of R.

Proposition 2.28 Let R be a subring of P, and let P be a subring of Q.

(a) If $R_R \leqq_r P_R$, then $P_R \leqq_r Q_R$ if and only if $P_P \leqq_r Q_P$.

(b) Q is a right quotient ring of R if and only if P is a right quotient ring of R and Q is a right quotient ring of P.

Proof: (a) If $P_R \leqq_r Q_R$ and $P_P \leqq M \leqq Q_P$, then $\mathrm{Hom}_R(M/P,Q) = 0$, whence $\mathrm{Hom}_P(M/P,Q) = 0$. Thus $P_P \leqq_r Q_P$.

Conversely, assume that $P_P \leqq_r Q_P$, and let $x,y \in Q$ with $x \neq 0$. First, there exists some $a \in P$ such that $xa \neq 0$ and $ya \in P$. Repeating this procedure with xa replacing both x and y, we find some $b \in P$ for which $xab \neq 0$ and $xab \in P$. Note that $yab \in P$ as well. Finally, since $R_R \leqq_r P_R$, there must be some $r \in R$ with $xabr \neq 0$ and $abr \in R$. Thus we have an element $s = abr \in R$ such that $xs \neq 0$ and $ys \in P$, which shows that $P_R \leqq_r Q_R$.

(b) follows automatically from (a), in view of 2.25.□

One reason we do not define a right quotient ring of R to be an overring Q such that $R_R \leqq_e Q_R$ is that transitivity of this condition [as in 2.28(b)] sometimes fails. (See Exercise 4.)

Definition A *maximal right quotient ring of* R is a right quotient ring Q such that whenever P is a right quotient ring of R, the inclusion map $R \to Q$ extends to an injective ring morphism $P \to Q$.

Theorem 2.29 Choose an injective hull $E = E(R_R)$ and set $T = \text{End}_R(E)$, $T_0 = \{f \in T \mid fR = 0\}$, $Q = \cap\{\ker f \mid f \in T_0\}$. Then Q has a unique ring structure compatible with its right R-module structure. Using this ring structure, Q is a maximal right quotient ring of R.

Proof: The module E is also a left T-module, and we set $W = \text{End}_T(E)$, writing maps in W on the *right* of their arguments in order to keep these maps on the opposite side from maps in T. We define **Z**-homomorphisms $T \to E$ and $W \to E$ by the rules $t \mapsto t1$ and $w \mapsto 1w$, and we claim that the first is an epimorphism and the second a monomorphism.

Given any $x \in E$, the map $r \mapsto xr$ of $R_R \to xR$ extends to a map $t \in T$ such that $t1 = x$. Thus $t \mapsto t1$ is an epimorphism. Now given any nonzero $w \in W$, we have $xw \neq 0$ for some $x \in E$, and consequently $(t1)w \neq 0$ for some $t \in T$. Inasmuch as $(t1)w = t(1w)$, we get $1w \neq 0$. Thus $w \mapsto 1w$ is a monomorphism.

For any $w \in W$, we have $f(1w) = (f1)w = 0$ for all $f \in T_0$, whence $1w \in Q$. Thus $w \mapsto 1w$ actually defines a **Z**-monomorphism $\phi : W \to Q$, and we claim that ϕ is an isomorphism. Given any $x \in Q$, we define a map $v : T \to E$ by the rule $tv = tx$. Note that $(T_0)v = 0$, because $x \in Q$. Inasmuch as $t \mapsto t1$ is a **Z**-epimorphism of T onto E with kernel T_0, we infer that v induces a **Z**-homomorphism $w : E \to E$ such that $(t1)w = tx$ for all $t \in T$, and clearly $w \in W$. Letting $j \in T$ denote the identity map, we observe that $\phi(w) = (j1)w = jx = x$. Therefore ϕ is an isomorphism.

With the help of ϕ, we can make Q into a ring by using the multiplication rule $x \cdot y = \phi[(\phi^{-1}x)(\phi^{-1}y)]$. Note that

$$x \cdot y = (1)(\phi^{-1}x)(\phi^{-1}y) = (x)(\phi^{-1}y)$$

Given any $r \in R$, right multiplication by r defines a map $w \in W$ such that $1w = r$, whence $\phi^{-1}r = w$. Thus for all $x \in Q$ we obtain $x \cdot r = xw = xr$. Therefore this ring structure on Q is compatible with its right R-module structure, and consequently we write xy for a product $x \cdot y$ in Q.

Now consider any ring multiplication $* : Q \times Q \to Q$ which is compatible with the right R-module structure. Given any $x \in Q$, we define a right R-module map $f : Q \to Q$ by the rule $fy = x*y - xy$, and we note that

$fR = 0$. Since $R_R \leqq_r Q_R$ by 2.26, we obtain $f = 0$. Therefore the ring structure on Q is unique.

We have already seen that $R_R \leqq_r Q_R$, whence Q is a right quotient ring of R. If P is any right quotient ring of R, then by 2.26 the inclusion map $R \to Q$ extends to a monomorphism $f : P_R \to Q_R$, and all that remains is to show that f is a ring morphism. Given any $x \in P$, we can define a map $g : (fP)_R \to Q_R$ by the rule $g(fy) = f(xy) - (fx)(fy)$, and we observe that $g(fR) = 0$. Inasmuch as $fR = R \leqq_r Q_R$, we obtain $g = 0$. Therefore f is indeed a ring morphism.□

Definition A *proper right quotient ring of* R is a right quotient ring Q such that R is a proper subring of Q.

Theorem 2.30 (a) If Q,Q' are both maximal right quotient rings of R, then the identity map on R extends to an isomorphism of Q onto Q'.

(b) A right quotient ring P of R is a maximal right quotient ring of R if and only if P has no proper right quotient rings.

Proof: (a) It suffices to consider the case when Q is the ring constructed in 2.29. In view of 2.26, we see that Q_R has no proper rational extensions. By definition, the inclusion map $R \to Q'$ extends to an injective ring map $f : Q \to Q'$. Observing that $Q_R \cong (fQ)_R$, we see that $(fQ)_R$ has no proper rational extensions, which implies that $fQ = Q'$. Therefore f is an isomorphism.

(b) If P is a maximal right quotient ring of R, then by (a) the identity map on R extends to an isomorphism of P onto the ring Q of 2.29. Observing that $P_R \cong Q_R$, we see by 2.26 that P_R has no proper rational extensions, whence 2.28 shows that P has no proper right quotient rings.

Conversely, assume that P has no proper right quotient rings, and let Q be a maximal right quotient ring of R. Then the inclusion map $R \to Q$ extends to an injective ring map $f : P \to Q$, and in view of 2.28 we see that Q is a right quotient ring of fP. Since fP is isomorphic to P and thus has no proper right quotient rings, we obtain $fP = Q$. Therefore f is a ring isomorphism of P onto Q, from which we conclude that P must be a maximal right quotient ring of R.□

Because of the uniqueness expressed in 2.30(a), we refer to *the* maximal right quotient ring of R whenever any maximal right quotient ring will suit our purposes.

The ring R also has a maximal *left* quotient ring, but in general the maximal right and left quotient rings of R need not even be isomorphic, let alone identical. (See Exercise 5.) In the next section, we derive necessary and sufficient conditions for a right and left nonsingular ring under which its maximal right and left quotient rings coincide.

Corollary 2.31 If $Z_r(R) = 0$, then $S^\circ R$ is a maximal right quotient ring of R. Thus the maximal right quotient ring of any right nonsingular ring is regular and right self-injective.

Proof: We have already observed that $S^\circ R$ is a right quotient ring of R. Now $S^\circ R$ is a regular, right self-injective ring by 2.11, and in particular, $(S^\circ R)_{S^\circ R}$ has no proper essential extensions. Consequently, it has no proper rational extensions either; hence $S^\circ R$ has no proper right quotient rings. According to 2.30, $S^\circ R$ is thus a maximal right quotient ring of R.□

Proposition 2.32 Let Q be a right quotient ring of R.

(a) $\mathscr{S}(Q) = \{I \leqq Q_Q \mid I \cap R \in \mathscr{S}(R)\}$ and $\mathscr{S}(R) = \{J \leqq R_R \mid JQ \in \mathscr{S}(Q)\}$.

(b) $Z_R(A) = Z_Q(A)$ for all A_Q.

(c) $Z_r(R) = R \cap Z_r(Q)$.

(d) $Z_r(R) = 0$ if and only if $Z_r(Q) = 0$.

Proof: (a) If $I \leqq Q_Q$ and $I \cap R \in \mathscr{S}(R)$, then $I \cap R \leqq_e R_R \leqq_r Q_R$ and consequently $I_R \leqq_e Q_R$. Then $I_Q \leqq_e Q_Q$ also.

Now let $I \in \mathscr{S}(Q)$ and consider any nonzero $x \in Q$. First, there must be some $a \in Q$ for which xa is a nonzero element of I. Since $R_R \leqq_r Q_R$, we also get an element $r \in R$ such that $xar \neq 0$ and $ar \in R$, whence $xR \cap I \neq 0$. Thus $I_R \leqq_e Q_R$, and consequently $I \cap R \in \mathscr{S}(R)$.

Given any $J \in \mathscr{S}(R)$, we have $JQ \cap R \in \mathscr{S}(R)$ also; hence $JQ \in \mathscr{S}(Q)$ by the above.

Finally, let $J \leqq R_R$ such that $JQ \in \mathscr{S}(Q)$, and consider any nonzero $x \in R$. According to the results above, $JQ \cap R \in \mathscr{S}(R)$; hence there must be some $r \in R$ such that xr is a nonzero element of $JQ \cap R$. Then $xr = y_1a_1 + \cdots + y_na_n$ for some $y_i \in J$, $a_i \in Q$. Inasmuch as $R_R \leqq_r Q_R$, there exist element $s_1, \ldots, s_n$ in R such that $xrs_1 \neq 0$ and $a_1s_1 \in R$, $xrs_1s_2 \neq 0$ and $a_2s_1s_2 \in R, \ldots,$ $xrs_1s_2 \cdots s_n \neq 0$ and $a_ns_1s_2 \cdots s_n \in R$. Setting $t = s_1s_2 \cdots s_n \in R$, we have $xrt \neq 0$ and $a_it \in R$ for all i. We now observe that xrt is a nonzero element of J, whence $xR \cap J \neq 0$. Therefore $J \in \mathscr{S}(R)$.

(b) and (c) follow directly from (a).

(d) Obviously $Z_r(Q) = 0$ implies $Z_r(R) = 0$. Conversely, if $Z_r(R) = 0$, we obtain $R \cap Z_r(Q) = 0$. Since $R_R \leqq_r Q_R$ and thus $R_R \leqq_e Q_R$, we conclude that $Z_r(Q) = 0$.□

Exercises

1. Find examples of modules A,B,C such that $A \leqq_r B$ but $A \oplus C \nleqq_r B \oplus C$.
2. Find examples of modules $A \leqq B \leqq C$ such that $B/A \leqq_r C/A$ but $B \nleqq_r C$.
3. Let F be a field, K a proper subfield of F, $R = \begin{pmatrix} F & 0 \\ F & K \end{pmatrix}$, $Q = \begin{pmatrix} F & F \\ F & F \end{pmatrix}$. Show that Q is a right quotient ring of R but not a left quotient ring.

4. Let F be a field, $T = F(x)[y]$, $Q = T/(y^2T)$, $P = (F[x] + F[x]y + y^2T)/(y^2T)$, $R = (F + F[x]y + y^2T)/(y^2T)$. Prove that $R_R \leqq_e P_R$ and $P_P \leqq_e Q_P$, but $R_R \nleqq_e Q_R$.
5. In Exercise 3, show that Q is the maximal right quotient ring of R, and prove that Q is not isomorphic to the maximal left quotient ring of R.
6. Given $A \leqq B$, show that $A \leqq_r B$ if and only if $\text{Hom}_R(B/A, E(B)) = 0$.
7. Show that R has no proper dense right ideals if and only if R_R contains a copy of every simple right R-module.
8. If R is simple, prime, or semiprime, show that every right quotient ring of R satisfies the same property.
9. Show that $Z_r(R) = 0$ if and only if R has a right quotient ring which is regular.
10. Assume that there is a ring decomposition $R = \Pi R_\alpha$. Prove that the maximal right quotient ring of R is the direct product of the maximal right quotient rings of the R_α.
11. Let F be a field, $T = F[x,y]$, $Q = T/(y^2T)$, $R = (F + F[x]y + y^2T)/(y^2T)$. Prove that $R_R \leqq_e Q_R$ but $R_R \nleqq_r Q_R$.
12. If R is commutative, prove that its maximal right and left quotient rings coincide and are commutative. Thus we omit the adjectives "right" and "left" in this case.
13. Let R be a commutative integral domain, K its quotient field. For any $n > 0$, prove that $K[x]/(x^n)$ is the maximal quotient ring of $R[x]/(x^n)$.
14. Let R be a commutative integral domain, K its quotient field, V a vector space over K. Make $Q = K \oplus V$ into a commutative ring as in Exercise 1.D.16, and show that Q is the maximal quotient ring of $R \oplus V$.
15. Let $\mathscr{S}_r(R)$ denote the set of all dense right ideals of R, and show that $\mathscr{S}_r(R)$ satisfies the properties listed in 1.19. Also, show that if $I \leqq R_R$, $J \in \mathscr{S}_r(R)$, and $r^{-1}I \in \mathscr{S}_r(R)$ for all $r \in J$, then $I \in \mathscr{S}_r(R)$.
16. Prove that $\mathscr{S}_r(R) = \mathscr{S}(R)$ if and only if $Z_r(R) = 0$.
17. With E,Q as in 2.29, show that $Q = \{x \in E \mid xI \leqq R \text{ for some } I \in \mathscr{S}_r(R)\}$.
18. If R is commutative, show that $\mathscr{S}_r(R)$ is exactly the set of those ideals of R which have zero annihilator.
19. Given any right R-module A, set $T_r(A) = \{x \in A \mid xI = 0 \text{ for some } I \in \mathscr{S}_r(R)\}$. Show that $T_r(A)$ is a submodule of A and that $T_r(A/T_r(A)) = 0$.
20. A right R-module A is *T_r-torsion* provided $T_r(A) = A$. Show that A is T_r-torsion if and only if $\text{Hom}_R(A, E(R_R)) = 0$. Conclude that the class of all T_r-torsion right R-modules is closed under factor modules, submodules, direct sums, and module extensions.
21. Given any right R-module A, choose an injective hull $E(A/T_r(A))$, and define

$$S_r^\circ A = \{x \in E(A/T_r(A)) \mid xI \leqq A/T_r(A) \text{ for some } I \in \mathscr{S}_r(R)\}$$

Show that $S_r^\circ R$ is the maximal right quotient ring of R. Prove that S_r° is an additive functor from right R-modules to right $S_r^\circ R$-modules.
22. Prove that S_r° is left exact.
23. If R has a minimal dense right ideal D, prove that S_r° is exact if and only if D is projective.
24. Let F be a field, R a vector space over F with basis $\{m_x \mid x \in [0,1)\}$, where $[0,1)$ is the half-open unit interval. Make R into a commutative F-algebra by

defining $m_x m_y = m_{x+y}$ when $x + y < 1$, $m_x m_y = 0$ when $x + y \geqq 1$. Show that $\mathscr{S}_r(R) = \{R, M\}$, where M is the (unique) maximal ideal of R, and prove that S_r° is not exact here.

D. Coincidence of Right and Left Quotient Rings

Given a ring R, we are interested in knowing when the maximal right and left quotient rings of R coincide, i.e., when every maximal right quotient ring of R is also a maximal left quotient ring, and vice versa. In view of the uniqueness of maximal quotient rings expressed in 2.30, this is equivalent to asking when there exists a ring which is both a maximal right quotient ring of R and a maximal left quotient ring of R. We give a complete answer to this question when R is right and left nonsingular. In this case all the maximal quotient rings involved are regular self-injective rings (by 2.31), and hence we begin with a number of results on regular self-injective rings. Our main tool is the supply of idempotents available in such rings. (Remember that every finitely generated one-sided ideal in a regular ring is generated by an idempotent.)

Lemma 2.33 Let e be an idempotent in a semiprime ring R. Then e is central if and only if eR is a two-sided ideal of R.

Proof: If eR is a two-sided ideal, then $Re \subseteq eR$ and so $(1 - e)Re = 0$. Since $R = Re + R(1 - e)$, it follows that $R(1 - e)$ is a two-sided ideal, hence $eR(1 - e)$ is a right ideal. Now $[eR(1 - e)]^2 = 0$, and R has no nonzero nilpotent right ideals, hence $eR(1 - e) = 0$. Given any $r \in R$, we thus have $er(1 - e) = 0$ as well as $(1 - e)re = 0$, whence $er = ere = re$. Therefore e is central. □

Definition Adopting the terminology used for operator algebras, we say that a regular ring R is *abelian* if all idempotents in R are central. Obviously commutative regular rings are abelian, but abelian regular rings need not be commutative. For example, any division ring is an abelian regular ring.

At the other extreme, we say that R is *totally non-abelian* if every nonzero right ideal of R contains a non-central idempotent. As Exercise 1 shows, this concept is left-right symmetric. For example, the ring of all $n \times n$ matrices over a division ring is totally non-abelian whenever $n \geqq 2$. (See also Exercise 2.)

Theorem 2.34 If R is a regular, right self-injective ring, then there is a ring decomposition $R = S \times T$ such that S is abelian and T is totally non-abelian.

Proof: Note from 1.27 that $Z_r(R) = 0$.

Let X denote the collection of those idempotents $e \in R$ for which every idempotent in eR is central (in R), and let S be the $\mathscr{S}$-closure of $(XR)_R$ in R_R. Now XR is a two-sided ideal because it is generated by central elements; hence S must be a two-sided ideal. Also, S_R is a closed submodule of the injective module R_R and so is injective; hence $S = sR$ for some idempotent s. According to 2.33, s must be central, which gives us a ring decomposition $R = S \times T$, where $T = (1 - s)R$.

Since T is a regular ring, any nonzero right ideal I of T must contain a nonzero idempotent e. Noting that $e \notin X$, we see that eR must contain an idempotent f which is not central in R. Certainly $f \in I$, and in particular $f \in T$. Then f commutes with everything in S, from which we infer that f is non-central in T. Therefore T is totally non-abelian.

We next claim that any nonzero right ideal J of S contains a nonzero central idempotent. Inasmuch as $XR \in \mathscr{S}(S)$ and $Z_r(S) = 0$, we have $JXR \neq 0$, whence $xe \neq 0$ for some $x \in J$, $e \in X$. By regularity, xeR contains a nonzero idempotent f. Since $f \in xeR \leqq eR$ and $e \in X$, f is central, and the claim is proved.

Now consider any idempotent $g \in S$, and let Y denote the collection of all central idempotents which lie in gS. We have just shown that every nonzero submodule of gS contains a nonzero element of Y, whence $YS \leqq_e gS$. Inasmuch as $S/gS \cong (1 - g)S$ is nonsingular, we infer from 2.3 that gS must be the $\mathscr{S}$-closure of YS in S_S. Now YS is two-sided because the idempotents in Y are central; hence gS must be two-sided, and finally 2.33 says that g is central. Therefore S is abelian. □

Theorem 2.35 Let R be an abelian regular ring. If Q is the maximal right quotient ring of R, then ${}_RQ$ is injective. Consequently, R is right self-injective if and only if it is left self-injective.

Proof: Since $Z({}_RQ)$ is a fully invariant submodule of ${}_RQ$, it must be a right ideal of Q. However, $R \cap Z({}_RQ) = Z_l(R) = 0$ (by 1.27); hence we obtain $Z({}_RQ) = 0$.

We also claim that any idempotent $e \in R$ must be central in Q. Otherwise, we have $xe - ex \neq 0$ for some $x \in Q$. Since $R_R \leqq_r Q_R$ there must be an element $r \in R$ such that $(xe - ex)r \neq 0$ and $xr \in R$. But then xr and r both commute with e (because R is abelian), which leads to the contradiction $(xe - ex)r = 0$.

Now let $I \leqq {}_RR$ and $f : I \rightarrow {}_RQ$. Since R is regular, I has an essential submodule of the form $\oplus Re_\alpha$, for suitable idempotents e_α. Also, since R is abelian, we have $Re_\alpha = e_\alpha R$ for all α; hence $\{e_\alpha R\}$ is an independent family of right ideals of R.

Set $x_\alpha = fe_\alpha$ for each α, and note that $x_\alpha = e_\alpha x_\alpha = x_\alpha e_\alpha$. For each α, left multiplication by x_α defines a map of $e_\alpha R$ into Q_R, and together these maps induce a map $g : \oplus e_\alpha R \to Q_R$. Inasmuch as $Q_R = (S^\circ R)_R$ is injective, g must be given by left multiplication by some $y \in Q$, that is, $x_\alpha e_\alpha = ye_\alpha$ for all α. Consequently, $x_\alpha = e_\alpha y$ for all α; hence right multiplication by y agrees with f on $\oplus Re_\alpha$. According to 2.1, f is given by right multiplication by y. Therefore $_RQ$ is injective.

Finally, if R_R is injective, we have $Q = R$ and so $_RR$ is injective by the above. The converse follows by symmetry.□

Lemma 2.36 Let R be regular and right self-injective. If R is totally non-abelian, then there exists a decomposition $R_R = I \oplus J \oplus K$ such that $I \cong J$ and K is isomorphic to a submodule of $I \oplus J$.

Proof: Let $\mathfrak{A}$ denote the collection of those right ideals A of R such that $A \cong B \oplus B$ for some B. Since $0 \in \mathfrak{A}$, $\mathfrak{A}$ is nonempty, and we can choose a maximal independent subfamily $\{A_\lambda\} \subseteq \mathfrak{A}$. For each λ, we have $A_\lambda = B_\lambda \oplus C_\lambda$ with $B_\lambda \cong C_\lambda$, which gives us a decomposition $\oplus A_\lambda = (\oplus B_\lambda) \oplus (\oplus C_\lambda)$. Inasmuch as R_R is injective, there exist right ideals $I = E(\oplus B_\lambda)$ and $J = E(\oplus C_\lambda)$, and since $\oplus B_\lambda \cong \oplus C_\lambda$ we obtain $I \cong J$. In view of 1.1, $I \cap J = 0$; hence by injectivity of $I \oplus J$ we get $R_R = I \oplus J \oplus K$ for some K. Inasmuch as $\oplus A_\lambda \leqq I \oplus J$, it follows from the maximality of $\{A_\lambda\}$ that 0 is the only member of $\mathfrak{A}$ which is contained in K.

Next let $\mathscr{F}$ denote the collection of all submodules of K which are isomorphic to submodules of $I \oplus J$. We claim that any nonzero $M \leqq K$ must contain a nonzero member of $\mathscr{F}$. Since R is totally non-abelian, M contains a non-central idempotent e. According to 2.33, eR is not a two-sided ideal of R; hence $Re \nsubseteq eR$ and so $(1 - e)Re \neq 0$. Thus there exists a nonzero map $f : eR \to (1 - e)R$. Since $f(eR)$ is a principal right ideal of R and therefore projective, $f(eR)$ must be isomorphic to some $B \leqq eR$. Now $B \cap f(eR) \leqq eR \cap (1 - e)R = 0$; hence $B \oplus f(eR) \in \mathfrak{A}$. Since $f(eR) \neq 0$, we cannot have $B \oplus f(eR) \leqq K$; hence the image of $B \oplus f(eR)$ under the projection $R \to I \oplus J$ is nonzero. Inasmuch as $B \cong f(eR)$, we infer from this that $\mathrm{Hom}_R(eR, I \oplus J) \neq 0$. Choosing a nonzero map $g : eR \to I \oplus J$, we have $g(eR)$ projective; hence $g(eR) \cong F$ for some $F \leqq eR \leqq M$. Since $F \neq 0$ and $F \cong g(eR) \leqq I \oplus J$, we have $F \in \mathscr{F}$, and the claim is proved.

Inasmuch as every nonzero submodule of K contains a nonzero member of $\mathscr{F}$, K must have an essential submodule of the form $\oplus F_\lambda$, where each $F_\lambda \in \mathscr{F}$. For each λ, we have a monomorphism $f_\lambda : F_\lambda \to I \oplus J$, and we claim that $\oplus f_\lambda : \oplus F_\lambda \to I \oplus J$ is a monomorphism also, i.e., that $\{f_\lambda F_\lambda\}$ is independent. If not, then there exist distinct indices $\lambda(0), \lambda(1), \ldots, \lambda(n)$ such that

$$f_{\lambda(0)}F_{\lambda(0)} \cap (f_{\lambda(1)}F_{\lambda(1)} + \cdots + f_{\lambda(n)}F_{\lambda(n)})$$

contains a nonzero element x. Since xR is projective, we obtain $xR \cong B \cong C$ for some $B \leqq F_{\lambda(0)}$ and some $C \leqq F_{\lambda(1)} \oplus \cdots \oplus F_{\lambda(n)}$. Then $B \cap C = 0$, so that we have $B \oplus C \in \mathfrak{A}$ and $B \oplus C \leqq K$. Since $B \oplus C \neq 0$, this is a contradiction. Therefore $\oplus f_\lambda$ is indeed a monomorphism.

Finally, since $I \oplus J$ is injective and $\oplus F_\lambda \leqq_e K$, the monomorphism $\oplus f_\lambda : \oplus F_\lambda \to I \oplus J$ must extend to a monomorphism $K \to I \oplus J$. Therefore K is isomorphic to a submodule of $I \oplus J$.□

Theorem 2.37 Let R be a regular, right self-injective ring. If R is totally non-abelian, then R is generated as a ring by its idempotents.

Proof: Let E denote the subring of R generated by its idempotents.

If $R_R = I \oplus J \oplus K$ as in 2.36, then R has orthogonal idempotents e_1, e_2, e_3 such that $e_1R = I$, $e_2R = J$, $e_3R = K$. Given any $r \in e_1Re_2$, we compute that $e_1 + r$ is an idempotent, whence $r = (e_1 + r) - e_1 \in E$. Thus $e_1Re_2 \subseteq E$, and similarly $e_iRe_j \subseteq E$ for any $i \neq j$.

According to 2.36, $e_1R \cong e_2R$. Since e_1, e_2 are idempotents, the isomorphism $e_1R \to e_2R$ is given by left multiplication by some $x \in e_2Re_1$, and its inverse by some $y \in e_1Re_2$. Note in particular that $yx = e_1$. Given any $r \in e_1Re_1$, we have $r = e_1r = yxr$ with $y \in e_1Re_2$ and $xr \in e_2Re_1$; hence it follows from the results above that $r \in E$. Thus $e_1Re_1 \subseteq E$, and similarly $e_2Re_2 \subseteq E$.

Now I $= e_1Re_1 + e_1Re_2 + e_1Re_3 \subseteq E$, and likewise $J \subseteq E$.

We also have $K \cong L$ for some $L \leqq I \oplus J$. Since K is a direct summand of R_R and thus injective, L is injective as well; hence we obtain $R_R = K \oplus L \oplus M$ for some M. Proceeding as above, we conclude that $K \subseteq E$. Therefore $R = I + J + K \subseteq E$.□

We are now in position to answer the question raised at the beginning of the section. The answer is in terms of annihilators, and we recall that a *right* (*left*) *annihilator ideal* in a ring R is any right (left) ideal which equals the right (left) annihilator of some subset of R. As is easily checked, a right ideal I is a right annihilator if and only if I is the right annihilator of the left annihilator of I, and similarly for left ideals.

Theorem 2.38 Let R be a right and left nonsingular ring. Then the maximal right and left quotient rings of R coincide if and only if every closed one-sided ideal of R is an annihilator.

Proof: The phrase "every closed one-sided ideal of R is an annihilator" means of course that every closed *right* ideal is a *right* annihilator, while every closed *left* ideal is a *left* annihilator.

According to 2.31, the maximal right (left) quotient ring of R is regular and right (left) self-injective, and we make use of these properties without further comment.

First assume that the maximal right and left quotient rings of R coincide in a ring Q, and consider any closed right ideal J of R. We have $Q = S^{\circ}R$ by 2.31, and $S^{\circ}J$ is an injective right ideal of Q by 2.9; hence $S^{\circ}J = eQ$ for some idempotent $e \in Q$. Since $J \leqq_e S^{\circ}J$ we obtain $J \leqq_e R \cap eQ$, and then $J = R \cap eQ$ (because J is closed in R_R). According to our hypotheses, we have ${}_RR \leqq_e {}_RQ$; hence $K = R \cap Q(1 - e)$ is an essential left R-submodule of $Q(1 - e)$. It now follows from 2.1 that the right annihilator of K *in* Q is the same as the right annihilator of $Q(1 - e)$, which is just eQ. Thus the right annihilator of K *in* R is exactly $R \cap eQ = J$. Therefore J is a right annihilator. By symmetry, closed left ideals of R are left annihilators also.

Conversely, assume that closed one-sided ideals of R are annihilators, and let Q be the maximal *right* quotient ring of R. We have $Z(Q_R) = 0$, and we obtain $Z({}_RQ) = 0$ as in 2.35.

Our main task is to show that Q is also a left quotient ring of R. We work with the largest subring P of Q which is a left quotient ring of R, first showing that P contains all the idempotents in Q, and using this to obtain $P = Q$.

Set P equal to the $\mathscr{S}$-closure of ${}_RR$ in ${}_RQ$. By 2.3, we see that ${}_RR \leqq_e {}_RP$. Given any $x \in P$, we observe that Px/Rx is an epimorphic image of the singular left R-module P/R, whence Px/Rx is singular. Since $Rx \leqq P$ and ${}_R(Q/P)$ is nonsingular, we must have $Px \leqq P$ as well. Therefore $P^2 \subseteq P$, so that P is a subring of Q. Since $Z_l(R) = 0$ and ${}_RR \leqq_e {}_RP$, P is now a left quotient ring of R.

According to 2.32, $Z_l(R) = 0$ implies $Z_l(P) = 0$. Also, 2.32 shows that $Z({}_P(Q/P)) = Z({}_R(Q/P)) = 0$, that is, ${}_PP$ is $\mathscr{S}$-closed in ${}_PQ$. On the other side, it follows from 2.28 and 2.30 that P is a right quotient ring of R, and that Q is the maximal right quotient ring of P. Consequently, one more application of 2.32 shows that $Z_r(P) = 0$.

Next, we show that any idempotent $e \in Q$ actually belongs to P. Since $R/(R \cap eQ)$ is isomorphic to a submodule of the nonsingular module $(1 - e)Q$, we obtain $R \cap eQ \in L^*(R_R)$. According to 2.4, $R \cap eQ$ is thus a closed right ideal of R; hence by hypothesis it must be a right annihilator. Inasmuch as $R \cap eQ \leqq_e eQ$, we infer from 2.1 that the left annihilator of $R \cap eQ$ is $R \cap Q(1 - e)$. Now $R \cap eQ$ must be the right annihilator of its left annihilator; hence $R \cap eQ$ is the right annihilator of $R \cap Q(1 - e)$. Likewise, $R \cap (1 - e)Q$ is the right annihilator of $R \cap Qe$. Setting $J = [R \cap Qe] + [R \cap Q(1 - e)]$, we thus see that the right annihilator of J is $[R \cap (1 - e)Q] \cap [R \cap eQ] = 0$. If K denotes the $\mathscr{S}$-closure of J in ${}_RR$, then the right annihilator of K is zero also. However, K is a closed left ideal of R and thus by hypothesis is an annihilator, hence we obtain $K = R$, that is, $J \leqq_e {}_RR$. Observing that $Je = R \cap Qe \leqq P$, we conclude that $e \in P$, as desired.

The first use of these idempotents is to show that P is regular. Given any $x \in P$, there is at least an element $y \in Q$ for which $xyx = x$. Replacing y by

yxy, we may assume that $yxy = y$ as well. Now $e = yx$ is an idempotent; hence $e \in P$, and $xy \in P$ for the same reason. Since $e \in P$, we see that the right annihilator of x *in* P is just $(1 - e)P$. Also, $R \cap Px \leqq_e Px$ (because ${}_RR \leqq_e {}_RP$); hence by using 2.1 we find that $(1 - e)P$ is also the right annihilator of $R \cap Px$ *in* P. Likewise, eP is the right annihilator of $R \cap P(1 - e)$ in P; hence we see that $J = [R \cap Px] + [R \cap P(1 - e)]$ has zero right annihilator *in* P. Consequently, the right annihilator of J *in* R is zero as well, and as in the last paragraph we infer from this that $J \leqq_e {}_RR$. Inasmuch as $y = ey$, we compute that $Jy = (R \cap Px)y \leqq Pxy \leqq P$, from which we conclude that $y \in P$. Thus P is regular, as claimed.

Now 2.34 gives us a ring decomposition $Q = Q_1 \times Q_2$ such that Q_1 is abelian and Q_2 is totally non-abelian. The units of Q_1 and Q_2 are idempotents and hence belong to P, so that we get a corresponding ring decomposition $P = P_1 \times P_2$ such that each Q_i is the maximal right quotient ring of P_i. According to 2.37, Q_2 is generated by its idempotents (all of which belong to P); hence we obtain $Q_2 = P_2$.

Since Q_1 is abelian, P_1 must be abelian too; hence 2.35 says that Q_1 is an injective left P_1-module. We know that ${}_PP$ is $\mathscr{S}$-closed in ${}_PQ$, from which we see that P_1 is a closed left P_1-submodule of Q_1; hence P_1 is left self-injective. Another application of 2.35 shows that P_1 is right self-injective, whence P_1 must be its own maximal right quotient ring, that is, $P_1 = Q_1$.

Thus we have $P = Q$, whence Q is a left quotient ring of R. If T is the maximal left quotient ring of Q, then by 2.28 and 2.30 we see that T is also the maximal left quotient ring of R. Inasmuch as our hypotheses on R are right-left symmetric, T must be a right quotient ring of R. But then T is a right quotient ring of Q (by 2.28); hence 2.30 says that $T = Q$. Therefore Q is also the maximal left quotient ring of R, and the proof is complete. □

Exercises

1. If R is regular, prove that every nonzero right ideal contains a non-central idempotent if and only if every nonzero left ideal contains a non-central idempotent.
2. If R is a prime, regular ring which is not a division ring, show that R is totally non-abelian.
3. Let $e = e^2 \in R$. Show that e is central if and only if e commutes with all idempotents in R.
4. If R is an abelian regular ring, show that all one-sided ideals of R are two-sided.
5. If R is regular, prove that R is abelian if and only if R/P is a division ring for all prime ideals P of R.
6. If R is regular and R/P is not a division ring for any prime ideal P of R, prove that R is totally non-abelian.

7. If R is regular, prove that R is totally non-abelian if and only if every nonzero right ideal contains a nonzero right ideal whose intersection with the center of R is zero.
8. Let R be regular and right self-injective. Prove that there is a ring decomposition $R = S \times T$ such that S_S has essential socle and T_T has zero socle.
9. Let $Z_r(R) = 0$. Prove that all closed *right* ideals of R are annihilators if and only if $\mathscr{S}(R)$ is exactly the set of those right ideals whose left annihilators are zero, if and only if every nonzero *left* ideal of $S^\circ R$ has nonzero intersection with R.
10. Let F be a field, V a finite-dimensional vector space over F, $R = \begin{pmatrix} F & 0 \\ V & F \end{pmatrix}$. Prove that the maximal right and left quotient rings of R are isomorphic, but that they coincide if and only if $[V : F] \leqq 1$. (*Hint*: Exercise 2.A.4.)
11. Let R be a commutative semiprime ring, and let T be the ring of all $n \times n$ matrices over R, for some positive integer n. Prove that every closed one-sided ideal of T is an annihilator.
12. If R is a ring whose maximal right and left quotient rings coincide, show that every right (left) quotient ring of R is also a left (right) quotient ring of R.
13. Let R be a prime ring with $\text{soc}(R_R) \neq 0$. Prove that the maximal right and left quotient rings of R coincide if and only if R is semisimple. (*Hint*: Exercise 1.C.17.)
14. Let R be regular and right self-injective. Prove that R is left self-injective if and only if all cyclic nonsingular left R-modules are projective.
15. Let R be regular, and let P be the set of all principal right ideals of R, ordered by inclusion. Show that P is a complemented modular lattice.
16. In Exercise 15, show that P is distributive if and only if R is abelian.
17. Let R be regular and either right or left self-injective. Prove that the lattice P of Exercise 15 is complete.
18. Let R be regular, and let B denote the set of all central idempotents in R. Show that B is a Boolean algebra under the operations $e \wedge f = ef$, $e \vee f = e + f - ef$, $e' = 1 - e$. If R is right self-injective, show that B is complete.
19. Let $Z_r(R) = 0$, and suppose that R contains an infinite independent sequence of nonzero pairwise isomorphic right ideals. Prove that $S^\circ R$ is not left self-injective. (*Hint*: Find a similar sequence in $S^\circ R$, and imitate 2.23.)
20. Let R be a regular, right and left self-injective ring. If $x, y \in R$ with $xy = 1$, prove that $yx = 1$ also. (*Hint*: Exercise 19.) Use Exercise 2.B.15 to extend this result to all right and left self-injective rings.

3

Finiteness Conditions

The purpose of this chapter is to study various conditions on a nonsingular ring R under which the localization S° and the maximal quotient ring $S^\circ R$ resemble the commutative localization discussed at the beginning of Section 2.A. To illustrate this more precisely, consider the relationship between a commutative integral domain R and its quotient field K. First, K consists of fractions; namely, every nonzero element of R is invertible in K, and every element of K has the form ab^{-1} for suitable $a,b \in R$. As a consequence of this, it is easily checked that K is a flat R-module, and that the map $K \otimes_R K \to K$ given by multiplication (that is, $x \otimes y \to xy$) is an isomorphism. In a similar fashion, it turns out that the commutative localization functor S^{-1} is naturally equivalent to the functor $(-) \otimes_R K$.

Section A is concerned with conditions on a right nonsingular ring R under which ${}_R(S^\circ R)$ is flat, and under which multiplication map $S^\circ R \otimes_R S^\circ R \to S^\circ R$ is an isomorphism. These conditions may be expressed as a requirement that certain right ideals of R have finitely generated essential submodules. Section B is concerned with the question of when the functors S° and $(-) \otimes_R S^\circ R$ are naturally equivalent. The answer here turns out to be the requirement that R_R does not contain any direct sums with infinitely many nonzero summands. This also happens to be the answer to the question of when $S^\circ R$ is a semisimple ring. Similar questions about the structure of $S^\circ R$ are dealt with in Section C, namely, when $S^\circ R$ is a division ring, and when $S^\circ R$ is a direct product of endomorphism rings of vector spaces. The answers here are requirements on the existence of right ideals of R which do not contain any direct sums with more than one nonzero summand. Finally, Section D considers the problem of when $S^\circ R$ can be expressed in terms of fractions of elements of R. The solutions here involve a combination of semiprimeness and the conditions studied in Section B.

A. Essential Finite Generation

Definition A module which has a finitely generated essential submodule is said to be *essentially finitely generated*. In particular, this includes all finitely generated modules. On the other hand, since $\mathbf{Z}_{\mathbf{Z}} \leqq_e \mathbf{Q}_{\mathbf{Z}}$, $\mathbf{Q}$ is an example of an essentially finitely generated $\mathbf{Z}$-module which is not finitely generated. (See also Exercises 1 and 2.) If I is an ideal in a commutative integral domain R, then either $I = 0$ or else $aR \leqq_e I$ for any nonzero $a \in I$, whence I is essentially finitely generated. This is another source for essentially finitely generated modules which are not finitely generated; for example, the ideal I generated by the indeterminates $x_1, x_2, \ldots$ in a polynomial ring $R = F[x_1, x_2, \ldots]$ in infinitely many variables over a field.

Definition A module A is said to be *essentially finitely related* provided there exists a short exact sequence $0 \to K \to F \to A \to 0$ such that F is finitely generated free and K is essentially finitely generated. This is satisfied in particular by all finitely presented modules. If I is any ideal in a commutative integral domain R, then we have seen that I is essentially finitely generated, whence R/I is essentially finitely related. (See also Exercise 3.) Since I can be chosen not to be finitely generated, we get examples of essentially finitely related modules R/I which are not finitely presented.

Proposition 3.1 (a) If B is a closed submodule of an essentially finitely generated module A, then A/B is essentially finitely generated.

(b) Let $B \leqq A$, and assume that $Z(A) = 0$. If B and A/B are both essentially finitely generated, then so is A.

(c) A direct sum $\oplus A_\alpha$ is essentially finitely generated if and only if all but finitely many $A_\alpha = 0$ and each A_α is essentially finitely generated.

Proof: (a) If M is a finitely generated essential submodule of A, then $M + B \leqq_e A$ as well; hence 1.4 says that $(M + B)/B \leqq_e A/B$. Thus $(M + B)/B$ is a finitely generated essential submodule of A/B.

(b) Let M be a finitely generated essential submodule of B, and choose a finitely generated submodule $K \leqq A$ such that $(K + B)/B \leqq_e A/B$. Note that $K + B \leqq_e A$. Now $(K + B)/(K + M)$ is an epimorphic image of the singular module B/M and so is singular; hence we obtain $K + M \leqq_e K + B$ from 1.21. Consequently, $K + M$ is a finitely generated essential submodule of A.

(c) First assume that all but finitely many $A_\alpha = 0$ and that each A_α is essentially finitely generated. Choose a finitely generated submodule $M_\alpha \leqq_e A_\alpha$ for each α, and note from 1.1 that $\oplus M_\alpha \leqq_e \oplus A_\alpha$. Since all but finitely many $M_\alpha = 0$, $\oplus M_\alpha$ is finitely generated; hence $\oplus A_\alpha$ is essentially finitely generated.

Conversely, assume that $\oplus A_\alpha$ has a finitely generated essential submodule M. There must exist indices $\alpha(1), \ldots, \alpha(n)$ such that $M \leqq A_{\alpha(1)} \oplus \cdots \oplus A_{\alpha(n)}$. For all $\alpha \notin \{\alpha(1), \ldots, \alpha(n)\}$, we thus obtain $M \cap A_\alpha = 0$, and consequently $A_\alpha = 0$. Since M is finitely generated, we can choose finitely generated submodules $M_\alpha \leqq A_\alpha$ for each α such that $M \leqq \oplus M_\alpha$. Then $\oplus M_\alpha \leqq_e \oplus A_\alpha$, from which we conclude that $M_\alpha \leqq_e A_\alpha$ for all α. Therefore each A_α is essentially finitely generated. □

With regard to 3.1(a), it is easily possible to have $B \leqq A$ so that A is essentially finitely generated but A/B is not (Exercise 4). Also, submodules of essentially finitely generated modules need not be essentially finitely generated (Exercise 5). The hypotheses in 3.1(b) can be relaxed somewhat (as in Exercise 6), but the hypothesis $Z(A) = 0$ cannot be eliminated entirely, as the following example shows. (See also Exercise 7.)

Example 3.2 There exist modules $B \leqq A$ such that B and A/B are both essentially finitely generated while A is not essentially finitely generated.

Proof: Set $R_n = \mathbf{Z}$ for $n = 1,2, \ldots, R = \Pi R_n, J = \oplus R_n$, and note that J is an ideal of R.

Let $B = (2R + J) \oplus 0 \leqq R \oplus (R/2J)$. Observing that $2R \leqq_e 2R + J$, we see that B is essentially finitely generated. Set $A = (1,\bar{1})R + B$, and note that A/B is cyclic, so that A/B is essentially finitely generated also.

If e_n denotes the unit of R_n, then $e_n \in J$ and so $(e_n,0) \in B$. We also have $(e_n,\bar{e}_n) = (1,\bar{1})e_n \in A$, whence $(0,\bar{e}_n) \in A$. Inasmuch as $(0,\bar{e}_n)$ is annihilated by $2e_nR \oplus (1 - e_n)R$, we infer that $(0,\bar{e}_n)R$ is a simple submodule of A.

Now suppose that F is a finitely generated essential submodule of A. Without loss of generality, we may assume that $(1,\bar{1}),(2,0) \in F$. Inasmuch as $A = (1,\bar{1})R + (2,0)R + (J \oplus 0)$, it follows that there exist elements $y_1, \ldots, y_k \in J$ such that F is generated by $(1,\bar{1}),(2,0),(y_1,0), \ldots, (y_k,0)$.

There are only finitely many of these y_i; hence we may choose a positive integer n such that $y_ie_n = 0$ for all i. Since $F \leqq_e A$, 1.16 shows that $(0,\bar{e}_n)R \leqq \operatorname{soc}(A) \leqq F$; hence we obtain

$$(0,\bar{e}_n) = (1,\bar{1})r + (2,0)s + (y_1,0)r_1 + \cdots + (y_k,0)r_k$$

for suitable $r,s,r_1, \ldots, r_k \in R$. Multiplying this equation by e_n, we find that

$$(0,\bar{e}_n) = (e_n,\bar{e}_n)r + (2e_n,0)s = (e_nr + 2e_ns,\bar{e}_nr)$$

As a result, $e_n r = -2e_n s$, whence $\bar{e}_n r = -2\bar{e}_n s = 0$. But $\bar{e}_n r = \bar{e}_n \neq 0$; hence we have a contradiction.

Therefore A is not essentially finitely generated. $\square$

Proposition 3.3 (a) Let $0 \to J \to P \to A \to 0$ be exact. If P is finitely generated projective and A is essentially finitely related, then J is essentially finitely generated.

(b) A finite direct sum $A_1 \oplus \cdots \oplus A_n$ is essentially finitely related if and only if each A_i is essentially finitely related.

(c) Let $0 \to C \to B \to A \to 0$ be exact, and assume that $Z(C) = 0$. If B is finitely generated and A is essentially finitely related, then C is essentially finitely generated.

(d) Let $0 \to C \to B \to A \to 0$ be exact, and assume that $Z_r(R) = 0$. If A and C are both essentially finitely related, then so is B.

Proof: (a) There exists an exact sequence $0 \to K \to F \to A \to 0$ with F finitely generated free and K essentially finitely generated. Using Schanuel's Lemma, we obtain $K \oplus P \cong J \oplus F$. Inasmuch as K and P are both essentially finitely generated, it now follows from 3.1 that J is essentially finitely generated.

(b) Since the A_i must be finitely generated in either case, we may choose exact sequences $0 \to K_i \to F_i \to A_i \to 0$ with each F_i finitely generated free. Inasmuch as the sequence $0 \to \oplus K_i \to \oplus F_i \to \oplus A_i \to 0$ is exact, we see by (a) that $\oplus A_i$ is essentially finitely related if and only if $\oplus K_i$ is essentially finitely generated. According to 3.1, this happens if and only if each K_i is essentially finitely generated, which by (a) is equivalent to having each A_i essentially finitely related.

(c) Choosing an epimorphism $f: F \to B$ with F finitely generated free, we construct a commutative diagram with exact rows and columns as follows:

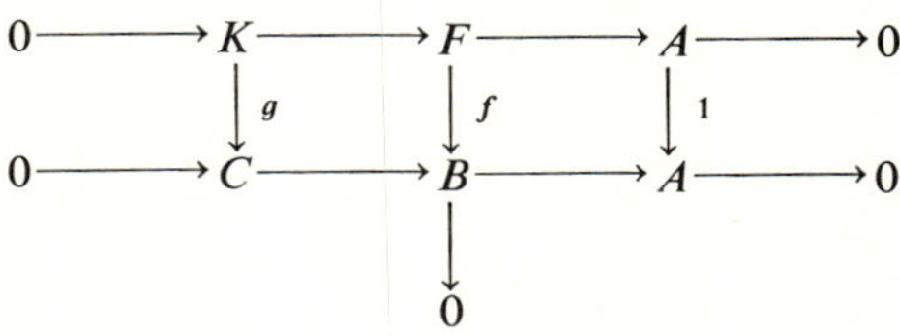

An easy diagram chase shows that g is surjective, whence $C \cong K/(\ker g)$. Since $Z(C) = 0$, we have $\ker g \in L^*(K)$, whence 2.4 shows that $\ker g$ is a closed submodule of K. Inasmuch as K is essentially finitely generated by (a), we conclude from 3.1 that C is essentially finitely generated also.

(d) Choosing exact sequences $0 \to K \to F \to A \to 0$ and $0 \to K' \to F' \to C \to 0$ with F, F' finitely generated free, we construct a commutative diagram with exact rows and columns as follows:

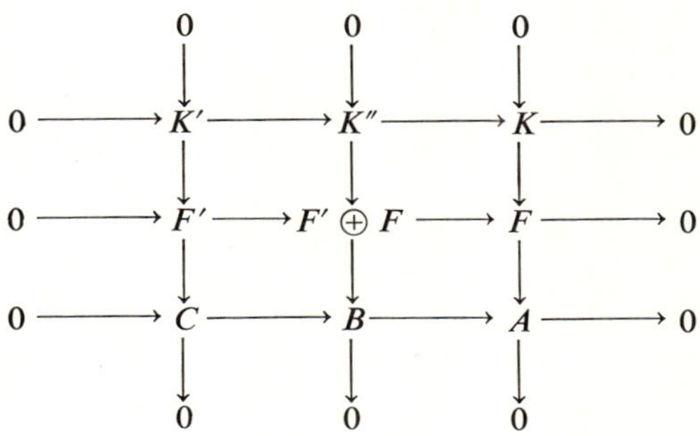

Inasmuch as $Z_r(R) = 0$, we have $Z(F' \oplus F) = 0$ and thus $Z(K'') = 0$. According to (a), K' and K are essentially finitely generated; hence it follows from 3.1 that K'' is essentially finitely generated. Since $F' \oplus F$ is finitely generated free, we conclude that B is essentially finitely related. □

Lemma 3.4 Let $Z_r(R) = 0$. For any right R-module A, we have a natural exact sequence $0 \to Z(A \otimes_R S^\circ R) \to A \otimes_R S^\circ R \to S^\circ A$.

Proof: The map $f : A \otimes_R S^\circ R \to S^\circ A$ is given by the rule $f(a \otimes x) = \bar{a}x$, where we are using $a \mapsto \bar{a}$ to denote the natural map $A \to A/Z(A) \leqq S^\circ A$. Inasmuch as $Z(S^\circ A) = 0$, we must have $Z(A \otimes_R S^\circ R) \leqq \ker f$. Conversely, if $\gamma = a_1 \otimes x_1 + \cdots + a_n \otimes x_n$ belongs to $\ker f$, then $\bar{a}_1 x_1 + \cdots + \bar{a}_n x_n = 0$. There exists a $J \in \mathscr{S}(R)$ such that $x_i J \leqq R$ for all i, and we observe that

$$\overline{a_1(x_1 r) + \cdots + a_n(x_n r)} = (\bar{a}_1 x_1 + \cdots + \bar{a}_n x_n) r = 0$$

for all $r \in J$. As a result, we see that $a_1(x_1 r) + \cdots + a_n(x_n r) \in Z(A)$ for all $r \in J$, and consequently

$$\gamma r = [a_1(x_1 r) + \cdots + a_n(x_n r)] \otimes 1 \in Z(A \otimes_R S^\circ R)$$

for all $r \in J$. Thus $\gamma J \leqq Z(A \otimes_R S^\circ R)$. According to 1.23, $(A \otimes_R S^\circ R)/Z(A \otimes_R S^\circ R)$ is nonsingular, from which we conclude that $\gamma \in Z(A \otimes_R S^\circ R)$. Therefore $Z(A \otimes_R S^\circ R) = \ker f$. □

Theorem 3.5 Let $Z_r(R) = 0$, and set $Q = S^\circ R$. Let A be a nonsingular right R-module, so that we have $AQ \leqq S^\circ A$.

(a) A is essentially finitely generated if and only if AQ is a finitely generated right Q-module.

(b) A is essentially finitely related if and only if A is finitely generated and $Z(A \otimes_R Q) = 0$.

Proof: (a) If A is essentially finitely generated, then it is clear from 2.10 that $(AQ)_Q$ is finitely generated. Conversely, if $(AQ)_Q$ is finitely generated, we must have $AQ = MQ$ for some finitely generated $M \leqq A_R$. Inasmuch as $M \leqq A \leqq MQ \leqq S°M$, we conclude that $M \leqq_e A$, whence A is essentially finitely generated.

(b) Since A is finitely generated in either case, we may choose an exact sequence $0 \to K \to F \to A \to 0$ with F finitely generated free. Applying the functors $(-)\otimes_R Q$ and $S°$, we obtain a commutative diagram with exact rows as follows:

$$\begin{array}{ccccccccc} & & K \otimes_R Q & \longrightarrow & F \otimes_R Q & \longrightarrow & A \otimes_R Q & \longrightarrow & 0 \\ & & \downarrow f & & \downarrow \cong & & \downarrow g & & \\ 0 & \longrightarrow & S°K & \longrightarrow & S°F & \longrightarrow & S°A & \longrightarrow & 0 \end{array}$$

Now $(S°F)_Q$ is finitely generated, and the bottom row splits because $(S°K)_Q$ is injective (by 2.9); hence $(S°K)_Q$ is finitely generated.

If A is essentially finitely related, then K is essentially finitely generated by 3.3; hence we obtain $S°K = KQ$ from 2.10. Since f is thus surjective, an easy diagram chase shows that g must be injective, whence $Z(A \otimes_R Q) = 0$ by 3.4.

Conversely, if $Z(A \otimes_R Q) = 0$, then 3.4 shows that g is injective. Reversing the diagram chase, we find that f is surjective, that is, $KQ = S°K$. As a result, $(KQ)_Q$ is finitely generated; hence K is essentially finitely generated by (a), and consequently A is essentially finitely related. □

Theorem 3.6 Let $Z_r(R) = 0$, and set $Q = S°R$. Then the following conditions are equivalent:

(a) ${}_RQ$ is flat.

(b) $Z(I \otimes_R Q) = 0$ for all right ideals I of R.

(c) Every finitely generated right ideal of R is essentially finitely related.

(d) If $x \in R$ and I is any finitely generated right ideal of R, then $\{r \in R \mid xr \in I\}$ is essentially finitely generated.

Proof: (a) ⇒ (b): If $I \leqq R_R$, then the natural map $I \otimes_R Q \to IQ$ is an isomorphism, from which we see that the natural map $I \otimes_R Q \to S°I$ is a monomorphism. By 3.4, $Z(I \otimes_R Q) = 0$.

(b) ⇒ (c) by 3.5.

(c) ⇒ (a): For any finitely generated right ideal I of R, we have $Z(I \otimes_R Q) = 0$ by 3.5; hence 3.4 shows that the map $I \otimes_R Q \to Q$ is injective. Therefore ${}_RQ$ is flat.

(c) ⇒ (d): If $J = \{r \in R \mid xr \in I\}$, then we have a short exact sequence $0 \to J \to R \oplus I \to xR + I \to 0$. Inasmuch as $xR + I$ is essentially finitely related by (c), we see by 3.3 that J must be essentially finitely generated.

(d) ⇒ (c): If $x \in R$ and $J = \{r \in R \mid xr = 0\}$, then $R/J \cong xR$. Since J is essentially finitely generated by (d), xR is essentially finitely related.

Now consider a right ideal $I = x_1R + \cdots + x_nR$ generated by $n > 1$ elements, and assume that $K = x_1R + \cdots + x_{n-1}R$ is essentially finitely related. Setting $J = \{r \in R \mid x_nr \in K\}$, we obtain $R/J \cong I/K$. Inasmuch as J is essentially finitely generated by (d), we see that I/K is essentially finitely related; hence we conclude from 3.3 that I must be essentially finitely related. □

In particular, 3.6(c) is satisfied by the following class of rings.

Definition A ring R is *right coherent* provided every finitely generated right ideal of R is finitely presented. For example, every right noetherian ring is right coherent. Also, since finitely generated projective modules are always finitely presented, all right semihereditary rings are right coherent. Thus every regular ring is both right and left coherent.

Corollary 3.7 If $Z_r(R) = 0$ and R is right coherent, then ${}_R(S^\circ R)$ is flat. □

Example 3.8 There exists a right nonsingular ring R such that neither ${}_R(S^\circ R)$ nor $(S^\circ R)_R$ is flat.

Proof: Let F be a field, z an indeterminate, $F_n = F(z)$ for $n = 0,1,2, \ldots, Q = \Pi F_n$. Set

$$J = \{x \in Q \mid x_0 \in zF[z] \text{ and } x_n = 0 \text{ for all but finitely many } n\}$$

Observing that J is an F-subspace of Q and that $J^2 \subseteq J$, we see that $R = F \cdot 1 + J$ is a subring of Q.

Since Q is a direct product of fields, it is regular and right self-injective. Noting that $R_R \leq_e Q_R$, we see from 2.11 that $Z_r(R) = 0$ and $S^\circ R = Q$.

Define $x \in R$ by setting $x_0 = z$ and $x_n = 0$ for all $n > 0$. Letting $A = \{r \in R \mid xr = 0\}$, we infer that $A = \bigoplus_{n=1}^{\infty} e_nR$, where e_n denotes the unit of F_n. In view of 3.1, we see that A is not essentially finitely generated; hence 3.6 shows that ${}_RQ$ is not flat. Since Q is commutative, Q_R is not flat either. □

Theorem 3.9 If R is a subring of a ring Q, then the following conditions are equivalent:

(a) ${}_RQ$ is flat, and the multiplication map $Q \otimes_R Q \to Q$ is an isomorphism.

(b) If $x \in Q$ and $J = \{r \in R \mid xr \in R\}$, then $JQ = Q$.

(c) The natural map $A \otimes_R Q \to Q$ is injective for all $A \leq Q_R$.

Proof: (c) ⇒ (a) is clear.

(a) $\Rightarrow$ (b): If j denotes the inclusion map $R_R \to Q_R$, then the composition of the map $j \otimes 1 : R \otimes_R Q \to Q \otimes_R Q$ with the isomorphism $Q \otimes_R Q \to Q$ yields the natural isomorphism $R \otimes_R Q \to Q$, from which we infer that $j \otimes 1$ must be an isomorphism. Since $j \otimes 1$ is surjective, we must have $(Q/R) \otimes_R Q = 0$. There is an obvious monomorphism $R/J \to (Q/R)_R$; hence in view of the flatness of $_RQ$ we obtain another monomorphism $(R/J) \otimes_R Q \to (Q/R) \otimes_R Q$. Thus $(R/J) \otimes_R Q = 0$, that is, $JQ = Q$.

(b) $\Rightarrow$ (c): For each $n \geqq 0$, let A_n denote the set of all elements of $A \otimes_R Q$ of the form $a_0 \otimes 1 + a_1 \otimes x_1 + \cdots + a_n \otimes x_n$, where $a_i \in A$, $x_i \in Q$. If f denotes the map $A \otimes_R Q \to Q$, we must show that $A_n \cap (\ker f) = 0$ for all n. It is clear that $A_0 \cap (\ker f) = 0$.

Now let $n > 0$ and assume that $A_{n-1} \cap (\ker f) = 0$. Consider any element

$$\gamma = a_0 \otimes 1 + a_1 \otimes x_1 + \cdots + a_n \otimes x_n \in A_n \cap (\ker f)$$

and set $J = \{r \in R \mid x_n r \in R\}$. For all $r \in J$, we have

$$\gamma r = [a_0 r + a_n(x_n r)] \otimes 1 + a_1 \otimes x_1 r + \cdots + a_{n-1} \otimes x_{n-1} r \in A_{n-1}$$

hence $\gamma J \leqq A_{n-1} \cap (\ker f) = 0$. Inasmuch as $JQ = Q$ by (b), we obtain $\gamma Q = 0$. Therefore $A_n \cap (\ker f) = 0$. □

Theorem 3.10 Let $Z_r(R) = 0$, and set $Q = S°R$. Then the following conditions are equivalent:

(a) $_RQ$ is flat, and the multiplication map $Q \otimes_R Q \to Q$ is an isomorphism.

(b) $Z(A \otimes_R Q) = 0$ for all nonsingular right R-modules A.

(c) Every finitely generated nonsingular right R-module is essentially finitely related.

(d) For all $x \in Q$, $\{r \in R \mid xr \in R\}$ is essentially finitely generated.

Proof: (a) $\Rightarrow$ (b): First note that

$$S°A \otimes_R Q \cong S°A \otimes_Q Q \otimes_R Q \cong S°A \otimes_Q Q \cong S°A$$

whence $Z(S°A \otimes_R Q) = 0$. Since $_RQ$ is flat, the monomorphism $A \to S°A$ induces another monomorphism $A \otimes_R Q \to S°A \otimes_R Q$, and consequently $Z(A \otimes_R Q) = 0$.

(b) $\Rightarrow$ (c) by 3.5.

(c) $\Rightarrow$ (d): If $J = \{r \in R \mid xr \in R\}$, then we obtain an exact sequence $0 \to J \to R \oplus R \to R + xR \to 0$. Now $R + xR$ is essentially finitely related by (c); hence 3.3 shows that J is essentially finitely generated.

(d) $\Rightarrow$ (a): For any $x \in Q$, the right ideal $J = \{r \in R \mid xr \in R\}$ belongs to $\mathscr{S}(R)$, from which we conclude that $S°J = Q$. Since J is essentially finitely generated by (d), 2.10 says that $JQ = Q$. Now (a) follows from 3.9. □

In particular, 3.10(d) is automatically satisfied in case R is right noetherian. The conditions of 3.10 obviously imply those of 3.6, but the following example shows that the reverse implication does not always hold.

Example 3.11 There exists a right nonsingular ring R such that ${}_R(S^\circ R)$ and $(S^\circ R)_R$ are both flat, but the multiplication map $S^\circ R \otimes_R S^\circ R \to S^\circ R$ is not an isomorphism.

Proof: Let F be a field, and set $F_n = F$ for $n = 1,2,\ldots$, $Q = \Pi F_n$, $J = \oplus F_n$. Since J is an ideal of Q, we see that $R = F\cdot 1 + J$ is a subring of Q. Now Q is a direct product of fields and so is regular and right self-injective. Observing that $R_R \leqq_e Q_R$, we see from 2.11 that $Z_r(R) = 0$ and $S^\circ R = Q$.

Note that R is exactly the set of those sequences $x = (x_1,x_2,\ldots)$ in Q which are eventually constant. Given $x \in R$, we can define $y \in Q$ by setting $y_n = 0$ whenever $x_n = 0$ and $y_n = x_n^{-1}$ whenever $x_n \neq 0$. Since x is eventually constant, so is y, whence $y \in R$. Also, $x_n y_n x_n = x_n$ for all n; hence $xyx = x$. Thus R is a regular ring. As a result, all R-modules are flat, in particular ${}_RQ$ and Q_R.

Now define $x \in Q$ by setting $x_n = 0$ for n odd and $x_n = 1$ for n even, and note that $xJ \leqq R$ but $x \notin R$. Since J is a maximal ideal of R, it follows that $J = \{r \in R \mid xr \in R\}$. Inasmuch as J is an infinite direct sum of nonzero ideals, 3.1 shows that J is not essentially finitely generated. Thus the multiplication map $Q \otimes_R Q \to Q$ cannot be an isomorphism, by 3.10. (See also Exercise 8.)□

The last part of this example is also a consequence of the following theorem, which is a rearrangement of 3.10 for the case of a regular ring.

Theorem 3.12 If R is regular, then the following conditions are equivalent:

(a) The multiplication map $S^\circ R \otimes_R S^\circ R \to S^\circ R$ is an isomorphism.

(b) R is right self-injective.

(c) All finitely generated nonsingular right R-modules are projective.

Proof: (a) ⇒ (b): As in the proof of 3.9, we obtain $(S^\circ R/R) \otimes_R S^\circ R = 0$. Inasmuch as R is regular, $(S^\circ R/R)_R$ is flat, hence we have a monomorphism $(S^\circ R/R) \otimes_R R \to (S^\circ R/R) \otimes_R S^\circ R$. Therefore $S^\circ R/R = 0$, whence $R = S^\circ R$ is right self-injective.

(b) ⇒ (c): According to 2.11, $S^\circ R = R$. If A is any finitely generated nonsingular right R-module, then $S^\circ A = A$ by 2.9, whence A is injective. Using 2.12, we can embed A in a finite direct sum F of copies of R_R. Since A is injective, it must be isomorphic to a direct summand of F, and thus is projective.

(c) ⇒ (a): Clearly all finitely generated nonsingular right R-modules are finitely presented; hence (a) follows from 3.10.□

Exercises

1. Let R be a commutative integral domain with quotient field K. Show that a torsion-free R-module A is essentially finitely generated if and only if $A \otimes_R K$ is finite-dimensional as a vector space over K.
2. Let A be a torsion **Z**-module, and set $A_p = \{x \in A \mid xp = 0\}$ for all primes p. Show that A is essentially finitely generated if and only if $A_p = 0$ for all but finitely many p and each A_p is finite-dimensional as a vector space over $\mathbf{Z}/p\mathbf{Z}$.
3. Prove that every finitely generated module over a commutative integral domain is essentially finitely related.
4. Let F be a field, $R = F[x_1, x_2, \ldots]$, A the ideal of R generated by the x_i, B the ideal generated by all products $x_i x_j$. Show that A is essentially finitely generated but that A/B is not.
5. Find an example of an essentially finitely generated module with a closed submodule which is not essentially finitely generated.
6. Let $Z(A) \leqq B \leqq A$. Prove that if B and A/B are both essentially finitely generated, then so is A.
7. Let R be right noetherian, and let A,B be right R-modules with $B \leqq A$. Prove that if B and A/B are both essentially finitely generated, then so is A.
8. Let Q,R be as in 3.11, and define $x,y \in Q$ by setting $x_n = 0$ and $y_n = 1$ for n odd, $x_n = 1$ and $y_n = 0$ for n even. Show that $x \otimes y$ is a nonzero element of the kernel of the multiplication map $Q \otimes_R Q \to Q$.
9. Show that an essentially finitely generated module cannot be the union of a chain of non-essential submodules.
10. If $Z_r(R) = 0$, prove that every essentially finitely generated projective right R-module is finitely generated.
11. Let Q,R be as in 3.11, and set $T = \begin{pmatrix} R & 0 \\ Q & Q \end{pmatrix}$. Show that $Z_r(T) = 0$ and $S^\circ T = \begin{pmatrix} Q & Q \\ Q & Q \end{pmatrix}$. Prove that ${}_T(S^\circ T)$ is not flat.
12. Let $Z_r(R) = 0$, and set $Q = S^\circ R$. Show that $Z[(Q \otimes_R Q)_R] = 0$ if and only if $Q \otimes_R (Q/R) = 0$, if and only if $(Q/R) \otimes_R Q = 0$.
13. Show that the conditions in 3.12 are also equivalent to having $(S^\circ R)_R$ projective.
14. If R is the ring of all lower triangular 2×2 matrices over a field F, prove that $Z_r(R) = 0$, and that ${}_R(S^\circ R)$ and $(S^\circ R)_R$ are both projective.
15. Let R be a commutative integral domain with quotient field K. Let $n > 0$, and let T,Q be the rings of all $n \times n$ matrices over R,K. Show that ${}_TQ$ is flat and that the multiplication map $Q \otimes_T Q \to Q$ is an isomorphism.
16. Let $\{R_\alpha\}$ be a collection of commutative integral domains, and for each α let K_α be the quotient field of R_α. If $R = \Pi R_\alpha$ and $Q = \Pi K_\alpha$, then $Z_r(R) = 0$ and $S^\circ R = Q$ (Exercise 2.A.12.) Show that ${}_RQ$ is flat and that $Q \otimes_R Q \to Q$ is an isomorphism.

17. Let F be a field. For $n = 1,2, \ldots$, let Q_n be the ring of all $n \times n$ matrices over F, and let R_n be the F-subalgebra of Q_n generated by 1 and the left-hand column of Q_n. According to Exercise 2.A.4, $Z_r(R_n) = 0$ and $S°R_n = Q_n$. Show that R_n satisfies the conditions of 3.10. If $R = \Pi R_n$ and $Q = \Pi Q_n$, then $Z_r(R) = 0$ and $S°R = Q$ (Exercise 2.A.12.) Prove that R does not satisfy the conditions of 3.10.

18. Let $n > 1$, and set $T = \mathbf{Q} \times (\mathbf{Z}/n\mathbf{Z})$. If R is the ring of all integer multiples of the identity in T, show that $T \otimes_R T \to T$ is an isomorphism, but that ${}_RT$ is not flat.

19. Let R be a subring of a ring Q such that $Q \otimes_R Q \to Q$ is an isomorphism. For any A_Q, prove that $A \otimes_R Q \to A$ and $A \to A \otimes_R Q$ are isomorphisms. For any A_Q, B_Q, prove that $\text{Hom}_R(A,B) = \text{Hom}_Q(A,B)$.

20. If R is a subring of a ring Q such that ${}_RQ$ is flat and $Q \otimes_R Q \to Q$ is an isomorphism, prove that Q is a right quotient ring of R.

21. If R is a subring of a ring Q such that ${}_RQ$ is flat and $Q \otimes_R Q \to Q$ is an isomorphism, prove that r.gl.gim.$(Q) \leqq$ r.gl.dim.(R).

B. Finite-dimensionality

Definition A module A is said to be *finite-dimensional* provided A contains no infinite independent families of nonzero submodules. Equivalently, A is finite-dimensional if and only if for any monomorphism $\oplus A_\alpha \to A$, all but finitely many $A_\alpha = 0$. When proving that A is finite-dimensional, it obviously suffices to show that A has no infinite independent *sequences* of nonzero submodules.

For example, all artinian modules and all noetherian modules are finite-dimensional. In particular, a vector space V over a division ring D is finite-dimensional in the sense just defined if and only if $[V : D] < \infty$; hence there is no conflict with the usual vector space terminology. Since $A \cap B \neq 0$ for all nonzero $\mathbf{Z}$-submodules $A,B \leqq \mathbf{Q}$, we see that $\mathbf{Q}$ is a finite-dimensional $\mathbf{Z}$-module which is neither artinian nor noetherian. Similarly, any ideal in a commutative integral domain is a finite-dimensional module. (See also Exercises 1 and 2.) Finally, note that any submodule of a finite-dimensional module is finite-dimensional.

Proposition 3.13 (a) A is finite-dimensional if and only if all submodules of A are essentially finitely generated.

(b) If A is finite-dimensional and B is closed in A, then A/B is finite-dimensional.

(c) Let $B \leqq A$. If B and A/B are both finite-dimensional, then so is A.

(d) If $A_1, \ldots, A_n$ are finite-dimensional, then so is $A_1 \oplus \cdots \oplus A_n$.

(e) If $A \leqq_e C$ and A is finite-dimensional, then so is C.

Proof: (a) If A is not finite-dimensional, then it has an infinite independent family $\{A_i\}$ of nonzero submodules. In this case, $\oplus A_i$ is a submodule of A which is not essentially finitely generated.

Now assume that A is finite-dimensional, and consider any $B \leqq A$. Choose a maximal independent family $\{C_i\}$ of nonzero cyclic submodules of B, and note that $\oplus C_i \leqq_e B$. Since A is finite-dimensional, the family $\{C_i\}$ must be finite, whence $\oplus C_i$ is finitely generated. Thus B is essentially finitely generated.

(b) Given $B \leqq M \leqq A$, we observe that B is also a closed submodule of M. Inasmuch as M is essentially finitely generated by (a), we see by 3.1 that M/B must be essentially finitely generated. It now follows from (a) that A/B is finite-dimensional.

(c) Consider any independent sequence $\{C_1,C_2, \ldots\}$ of submodules of A. We first claim that $B \cap \left(\bigoplus_{n=k}^{\infty} C_n \right) = 0$ for some k. If not, we construct a sequence $\{B_1,B_2, \ldots\}$ of submodules of B as follows. First set $n_1 = 0$, and choose a nonzero element $b_1 \in B \cap \left(\bigoplus_{n=1}^{\infty} C_n \right)$. Then $b_1 \in \bigoplus_{n=1}^{n_2} C_n$ for some $n_2 > 0$, and we set $B_1 = B \cap \left(\bigoplus_{n=1}^{n_2} C_n \right) \neq 0$. Next, $B \cap \left(\bigoplus_{n=n_2+1}^{\infty} C_n \right) \neq 0$; hence we must have $B_2 = B \cap \left(\bigoplus_{n=n_2+1}^{n_3} C_n \right) \neq 0$ for some $n_3 > n_2$. Continuing in this manner, we obtain a sequence of integers $n_1 < n_2 < \cdots$ such that $B_i = B \cap \left(\bigoplus_{n=n_i+1}^{n_{i+1}} C_n \right)$ is nonzero for all i. But then $\{B_1,B_2, \ldots\}$ is an infinite independent sequence of nonzero submodules of B, which is impossible.

Therefore we must have $B \cap \left(\bigoplus_{n=k}^{\infty} C_n \right) = 0$ for some k. Consequently, there is a monomorphism of $\bigoplus_{n=k}^{\infty} C_n$ into the finite-dimensional module A/B; hence there must be an integer $N \geqq k$ such that $C_n = 0$ for all $n \geqq N$. Thus A is finite-dimensional.

(d) follows from (c) by induction.

(e) If $\{C_i\}$ is an independent family of submodules of C, then $\{A \cap C_i\}$ is an independent family of submodules of A, hence we obtain $A \cap C_i = 0$ for all but finitely many i. Inasmuch as $A \leqq_e C$, it follows that $C_i = 0$ for all but finitely many i. Therefore C is finite-dimensional. □

We note that 3.13(b) may fail if B is not closed in A (Exercise 3). In view of 3.13(d), we see that R_R is finite-dimensional if and only if all finitely generated projective right R-modules are finite-dimensional. Since these latter concepts can be expressed in categorical terms, we see that if R and S are Morita-equivalent rings, then R_R is finite-dimensional if and only if S_S is finite-dimensional. In particular, finite-dimensionality carries over from R to the ring of all $n \times n$ matrices over R, for any n. A much harder problem is to show that finite-dimensionality carries over from R to polynomial rings over R. The proof requires the use of uniform modules, which we study in the next section.

Theorem 3.14 For any module A, the following conditions are equivalent:

(a) A is finite-dimensional.

(b) A has ACC on closed submodules.

(c) A has DCC on closed submodules.

Proof: (a) $\Rightarrow$ (b): If not, then A has a chain $A_1 < A_2 < \cdots$ of closed submodules. For each n, A_n is a proper closed submodule of A_{n+1}; hence $A_n \not\leqq_e A_{n+1}$. Thus A_{n+1} must have a nonzero submodule C_n such that $A_n \cap C_n = 0$. For all n,

$$(C_1 + \cdots + C_n) \cap C_{n+1} \leqq A_{n+1} \cap C_{n+1} = 0$$

hence $\{C_1,C_2,\ldots\}$ is an independent sequence of nonzero submodules of A, which is impossible.

(b) $\Rightarrow$ (c): If not, then A has a chain $A_1 > A_2 > \cdots$ of closed submodules. Let B_1 be a relative complement for A_1 in A. Since $B_1 \cap A_2 \leqq B_1 \cap A_1 = 0$, we can enlarge B_1 to a relative complement B_2 for A_2. Continuing in this manner, we obtain a chain $B_1 \leqq B_2 \leqq \cdots$ such that each B_n is a relative complement for A_n in A. According to 1.4, each B_n is closed in A; hence it follows from (b) that $B_n = B_{n+1}$ for some n. Now $A_{n+1} \oplus B_{n+1} \leqq_e A$ by 1.3; hence $A_{n+1} \oplus B_n \leqq_e A$. Inasmuch as $A_{n+1} \oplus B_n \leqq A_n \oplus B_n$, it follows that $A_{n+1} \leqq_e A_n$. Since A_{n+1} is closed in A, we obtain $A_n = A_{n+1}$, which is false.

(c) $\Rightarrow$ (a): Consider any independent sequence $\{C_1,C_2,\ldots\}$ of submodules of A, and set $A_0 = A$. Since $C_1 \cap \left(\bigoplus_{n=2}^{\infty} C_n\right) = 0$, we can enlarge $\bigoplus_{n=2}^{\infty} C_n$ to a relative complement A_1 for C_1 in A_0. Continuing in this manner, we obtain a chain $A_1 \geqq A_2 \geqq \cdots$ such that for all $k > 0$, $\bigoplus_{n=k+1}^{\infty} C_n \leqq A_k$ and A_k is a relative complement for C_k in A_{k-1}. Then A_k is closed in A_{k-1} for all k (by 1.4); hence it follows from 1.5 that each A_k is closed in A. In view of (c), there must be a positive integer N such that $A_n = A_N$ for all $n \geqq N$; hence

$$C_{n+1} = C_{n+1} \cap A_n = C_{n+1} \cap A_{n+1} = 0$$

for all $n \geqq N$. Therefore A is finite-dimensional. $\square$

For a nonsingular module A, 2.4 says that the set of closed submodules of A is just $L^*(A)$; hence 3.14 shows that A is finite-dimensional if and only if $L^*(A)$ has either ACC or DCC. This can be used to relate finite-dimensionality of A to the structure of $S^\circ A$, but first some results about $L^*(A)$ and $L^*(S^\circ A)$ are needed.

Lemma 3.15 Let $Z_r(R) = 0$, and let A be any right R-module.

(a) $K \mapsto K/Z(A)$ defines a lattice isomorphism of $L^*(A)$ onto $L^*(A/Z(A))$.

(b) $K \mapsto S^\circ K$ defines a lattice isomorphism of $L^*(A)$ onto $L^*(S^\circ A)$.

Proof: (a) Since $Z(A)$ is the $\mathscr{S}$-closure of 0 in A (by 2.3), it is the smallest element of $L^*(A)$. Whenever $Z(A) \leqq K \leqq A$, we have $A/K \cong [A/Z(A)]/[K/Z(A)]$, from which we infer that $K \in L^*(A)$ if and only if $K/Z(A) \in L^*(A/Z(A))$. Therefore the rule $K \mapsto K/Z(A)$ defines a bijection θ of $L^*(A)$ onto $L^*(A/Z(A))$. For $K,K' \in L^*(A)$, we note that $K \leqq K'$ if and only if $\theta(K) \leqq \theta(K')$, whence θ is a lattice isomorphism.

(b) Whenever $Z(A) \leqq K \leqq A$, 2.5 says that $S^\circ K$ is the $\mathscr{S}$-closure of $K/Z(A)$ in $S^\circ A$; hence $S^\circ K \in L^*(S^\circ A)$. Thus the rule $K \mapsto S^\circ K$ defines an order-preserving map $\phi : L^*(A) \to L^*(S^\circ A)$. If $f : A \to S^\circ A$ denotes the natural map, then for any $M \in L^*(S^\circ A)$ we have a monomorphism of $A/f^{-1}M$ into the nonsingular module $S^\circ A/M$, from which we see that $f^{-1}M \in L^*(A)$. Therefore we can define an order-preserving map $\psi : L^*(S^\circ A) \to L^*(A)$ by the rule $\psi(M) = f^{-1}M$.

For any $K \in L^*(A)$, we have $K/Z(A) \leqq_e S^\circ K \cap [A/Z(A)]$. Inasmuch as $K/Z(A) \in L^*(A/Z(A))$ by (a), we obtain $S^\circ K \cap [A/Z(A)] = K/Z(A)$, from which it follows that $f^{-1}(S^\circ K) = K$, that is, $\psi\phi(K) = K$. On the other hand, for any $M \in L^*(S^\circ A)$ we see that $ff^{-1}M = M \cap [A/Z(A)] \leqq_e M$, whence $M/ff^{-1}M$ is singular. According to 2.3, M is the $\mathscr{S}$-closure of $ff^{-1}M$ in $S^\circ A$, and consequently 2.5 shows that $M = S^\circ(f^{-1}M)$, that is, $\phi\psi(M) = M$.

Therefore ϕ and ψ are inverse order-preserving maps; hence they are lattice isomorphisms. □

Theorem 3.16 Let $Z_r(R) = 0$. For any right R-module A, the following conditions are equivalent:

(a) $(S^\circ A)_{S^\circ R}$ is noetherian.

(b) $(S^\circ A)_{S^\circ R}$ is finitely generated semisimple.

(c) $(S^\circ A)_{S^\circ R}$ is artinian.

(d) $A/Z(A)$ is finite-dimensional.

(e) $L^*(A)$ has ACC.

(f) $L^*(A)$ has DCC.

Proof: Inasmuch as $Z(A/Z(A)) = 0$ by 1.23, 2.4 says that the set of closed submodules of $A/Z(A)$ is exactly $L^*(A/Z(A))$. Since $L^*(A/Z(A)) \cong L^*(A)$ by 3.15, we see by 3.14 that (d) $\Leftrightarrow$ (e) $\Leftrightarrow$ (f).

(a) ⇒ (b): Since all finitely generated submodules of $(S^\circ A)_{S^\circ R}$ are direct summands of $S^\circ A$ (by 2.9), this is clear.

(b) ⇒ (c) is clear.

(c) ⇒ (f): In view of 2.9, we see that $L^*(S^\circ A)$ must have DCC, hence 3.15 shows that $L^*(A)$ has DCC.

(e) ⇒ (a): According to 3.15, $L^*(S^\circ A)$ must have ACC. In view of 2.9, we see that $(S^\circ A)_{S^\circ R}$ has ACC on finitely generated submodules, from which we conclude that $(S^\circ A)_{S^\circ R}$ is noetherian.□

Definition A *cofinal subset* of $\mathscr{S}(R)$ is a subset $\mathscr{S}'$ such that every member of $\mathscr{S}(R)$ contains a member of $\mathscr{S}'$.

Theorem 3.17 If $Z_r(R) = 0$, then the following conditions are equivalent:

(a) The functors S° and $(-) \otimes_R S^\circ R$ are naturally equivalent.

(b) $S^\circ A = A$ for all right $S^\circ R$-modules A.

(c) $S^\circ R$ is a semisimple ring.

(d) R_R is finite-dimensional.

(e) $I(S^\circ R) = S^\circ R$ for all $I \in \mathscr{S}(R)$.

(f) $\mathscr{S}(R)$ has a cofinal subset of finitely generated right ideals.

(g) All direct sums of nonsingular injective right R-modules are injective.

Proof: (c) ⇔ (d) by 3.16.

(c) ⇒ (b): Since A is a projective right $S^\circ R$-module and $Z(S^\circ R) = 0$, we infer that $Z(A) = 0$. Similarly, $Z(S^\circ A/A) = 0$, and consequently $S^\circ A/A = 0$.

(b) ⇒ (a): For any A_R, we have the natural map $m_A : A \otimes_R S^\circ R \to S^\circ A$ given by the rule $m_A(a \otimes x) = ax$. Inasmuch as $S^\circ(A \otimes_R S^\circ R) = A \otimes_R S^\circ R$ by (b), we see that $Z(A \otimes_R S^\circ R) = 0$, and then 3.4 says that m_A is injective. Using (b) again, we obtain $S^\circ[m_A(A \otimes_R S^\circ R)] = m_A(A \otimes_R S^\circ R)$, from which we conclude that $m_A(A \otimes_R S^\circ R) = S^\circ A$. Thus the maps m_A are all isomorphisms; hence m is a natural equivalence between $(-) \otimes_R S^\circ R$ and S°.

(a) ⇒ (e): According to (a), $(R/I) \otimes_R S^\circ R \cong S^\circ(R/I)$, which is zero because R/I is singular. Thus $[S^\circ R]/[I(S^\circ R)] = 0$.

(e) ⇒ (f): We see by (e) that $I(S^\circ R)$ is a cyclic right $S^\circ R$-module for all $I \in \mathscr{S}(R)$; hence 3.5 shows that every $I \in \mathscr{S}(R)$ is essentially finitely generated, which is equivalent to (f).

(f) ⇒ (d): According to (f), every $I \in \mathscr{S}(R)$ is essentially finitely generated. Inasmuch as every right ideal of R is a direct summand of a member of $\mathscr{S}(R)$, it follows from 3.1 that all right ideals of R are essentially finitely generated. By 3.13, R_R is finite-dimensional.

(b) ⇒ (g): If $\{E_\alpha\}$ is any collection of nonsingular injective right R-modules, then $S^\circ E_\alpha = E_\alpha$ for all α, whence each E_α is a right $S^\circ R$-module. According to (b), $S^\circ(\oplus E_\alpha) = \oplus E_\alpha$, whence $\oplus E_\alpha$ is injective.

(g) $\Rightarrow$ (c): Applying 3.16 to the module $A = (S^\circ R)_{S^\circ R}$, we see that it suffices to show that A is finite-dimensional. If not, then A contains an infinite independent sequence $\{C_n\}$ of nonzero cyclic $S^\circ R$-submodules. According to 2.9, $S^\circ C_n = C_n$ for all n, hence each C_n is a nonsingular injective right R-module. By (g), $\oplus C_n$ is injective as a right R-module, whence $S^\circ(\oplus C_n) = \oplus C_n$. Then $\oplus C_n$ is also injective as a right $S^\circ R$-module, by 2.9. As a result, $\oplus C_n$ is a direct summand of A and so is a cyclic right $S^\circ R$-module, which is impossible. Therefore A is finite-dimensional, and consequently $S^\circ R$ is semisimple. □

For additional conditions equivalent to those given in 3.17, see Exercises 4 to 6. In view of 3.13, we see that 3.17(d) implies 3.10(d); hence the conditions of 3.17 imply those of 3.10. The converse implication, however, fails. For example, if R is the direct product of an infinite family of fields, then $Z_r(R) = 0$ and $S^\circ R = R$; hence 3.10(a) is trivially satisfied. On the other hand, 3.17(c) fails because R is not a semisimple ring.

Exercises

1. Let R be a commutative integral domain with quotient field K. Show that a torsion-free R-module A is finite-dimensional if and only if $A \otimes_R K$ is finite-dimensional as a vector space over K.
2. Show that a torsion $\mathbf{Z}$-module A is finite-dimensional if and only if soc(A) is finitely generated.
3. With A,B as in Exercise 3.A.4, show that A is finite-dimensional but that A/B is not.
4. If $Z_r(R) = 0$, prove that R_R is finite-dimensional if and only if every nonsingular right R-module has a largest injective submodule.
5. If $Z_r(R) = 0$, prove that R_R is finite-dimensional if and only if every nonsingular injective right R-module is a direct sum of indecomposable modules.
6. If $Z_r(R) = 0$, prove that R_R is finite-dimensional if and only if the class of nonsingular right R-modules is closed under direct limits.
7. Let $Z_r(R) = 0$ and assume that R_R is finite-dimensional. Prove that a nonsingular right R-module A is finite-dimensional if and only if A is essentially finitely generated.
8. If R is right noetherian, prove that a right R-module A is finite-dimensional if and only if A is essentially finitely generated.
9. Prove that a module A is finite-dimensional if and only if $E(A)$ is a finite direct sum of indecomposable modules.
10. Let R be commutative and nonsingular. Prove that R is finite-dimensional if and only if every prime ideal of R is essentially finitely generated.
11. Let R be commutative and nonsingular. Prove that R is finite-dimensional if and only if R has only finitely many minimal prime ideals.
12. Prove that every finite-dimensional module is a direct sum of indecomposable modules.

13. Prove that a module A is finitely generated semisimple if and only if A is finite-dimensional and all cyclic submodules of A are direct summands of A.

14. Let R be regular, and let I be a finitely generated right ideal of R. Prove that I is semisimple if and only if it is finite-dimensional.

15. Prove that a commutative ring R is a finite direct product of integral domains if and only if R is finite-dimensional and all principal ideals of R are projective.

16. Let Q be a right quotient ring of R. If I is a right ideal of R, prove that I_R is finite-dimensional if and only if $(IQ)_Q$ is finite-dimensional.

17. Let $Z_r(R) = 0$ and assume that R_R is finite-dimensional. If $\{A_\alpha\}$ is any collection of nonsingular right R-modules, show that $\bigoplus A_\alpha \in L^*(\Pi A_\alpha)$.

18. If $Z_r(R) = 0$ and R_R is finite-dimensional, show that $L^*(R_R)$ is isomorphic to the lattice of right ideals of $S^\circ R$.

19. Let $Z_r(R) = 0$ and assume that R_R is finite-dimensional. If A is a nonsingular right R-module which has a countably generated essential submodule, show that A can be embedded in a countable direct sum of copies of $S^\circ R$.

20. Let $Z_r(R) = 0$ and assume that R_R is finite-dimensional. Prove that R has a two-sided ideal J such that J_R is injective and J contains all injective right ideals of R.

21. Let F be a field, and let R be the ring of all $n \times n$ lower triangular matrices over F, for some $n > 1$. Show that $Z_r(R) = 0$ and that R_R is finite-dimensional. Show that the ideal J of Exercise 20 is the bottom row of R.

22. If R is semiprime and R_R is finite-dimensional, show that there is a ring decomposition $R = S \times T$ such that S is semisimple and $\mathrm{soc}(T_T) = 0$. (*Hint*: Exercise 1.C.17.)

23. Let $Z_r(R) = 0$ and assume that R_R is finite-dimensional. Prove that the maximal right and left quotient rings of R coincide if and only if $(S^\circ R)_R$ is flat. (*Hint*: Exercise 3.A.20.)

C. Uniform Modules

For a module A to be finite-dimensional, we require that all independent families of nonzero submodules of A are finite, but we do not assume there to be a bound on the cardinalities of these families. Actually, such a bound always exists, as we prove shortly. In order to prove this, we first study the simplest case: A module A such that independent families of nonzero submodules of A have at most one member.

Definition A *uniform module* is a nonzero module A such that any two nonzero submodules of A have nonzero intersection. Equivalently, A is uniform if and only if $A \neq 0$ and every nonzero submodule of A is essential in A. For example, every simple module is uniform. Also, **Q** is a uniform **Z**-module, and any nonzero ideal in a commutative integral domain is a uniform module. (See also Exercises 1 and 2.) Note that any nonzero submodule of a uniform module is uniform, and that any essential extension of a uniform module is uniform. On the other hand, factor modules of uniform modules need not be uniform (Exercise 3).

Lemma 3.18 If A is finite-dimensional, then every nonzero submodule of A contains a uniform submodule.

Proof: If B is a nonzero submodule of A, then B is finite-dimensional and so has DCC on closed submodules, by 3.14. Since B already has at least two distinct closed submodules, namely, 0 and B, we infer that B must have a minimal nonzero closed submodule K. If K is not uniform, then it has nonzero submodules C,D such that $C \cap D = 0$. Enlarging D if necessary, we may assume that D is a relative complement for C in K. Then D is closed in K by 1.4 and thus closed in B, by 1.5. However, $D \neq 0$ and $D \neq K$ (because $C \neq 0$), which contradicts the minimality of K. Therefore K is a uniform submodule of B.□

In order to include the zero module in what follows, we must allow for the direct sum of an empty family of uniform modules. By convention, we write this as $A_1 \oplus \cdots \oplus A_0$, so that the expression $A_1 \oplus \cdots \oplus A_n$ covers the case $n = 0$.

Proposition 3.19 (a) A module A is finite-dimensional if and only if A has an essential submodule which is a direct sum of finitely many uniform submodules.

(b) Assume that $A_1 \oplus \cdots \oplus A_n \leqq_e A$ with each A_i uniform. Then any independent family of nonzero submodules of A has at most n members.

Proof: (a) First suppose that $A_1 \oplus \cdots \oplus A_n \leqq_e A$ with each A_i uniform. Since each A_i is finite-dimensional, it follows from 3.13 that A is finite-dimensional.

Conversely, let A be finite-dimensional, and choose a maximal independent family $\{A_i\}$ of uniform submodules of A. Since every nonzero submodule of A contains a uniform submodule (by 3.18), it follows from the maximality of $\{A_i\}$ that $\oplus A_i \leqq_e A$. Also, since A is finite-dimensional, the family $\{A_i\}$ must be finite.

(b) If $n = 0$, then $0 \leqq_e A$ and $A = 0$, in which case A has no nonzero submodules. If $n = 1$, then $A_1 \leqq_e A$ and so A is uniform, in which case independent families of nonzero submodules of A can have at most one member.

Now let $n > 1$, and assume that (b) holds for modules which have an essential submodule consisting of a direct sum of $n - 1$ uniform submodules.

Suppose that $B_1 \oplus \cdots \oplus B_{n+1} \leqq A$ with each $B_i \neq 0$. If $B = B_1 \oplus \cdots \oplus B_n$, then $B \cap B_{n+1} = 0$ and so $B \nleqq_e A$, from which we infer that $(B \cap A_1) \oplus \cdots \oplus (B \cap A_n) \nleqq_e A$. Inasmuch as $A_1 \oplus \cdots \oplus A_n \leqq_e A$, it follows that $B \cap A_i \nleqq_e A_i$ for some i, and after renumbering we may assume that $B \cap A_n \nleqq_e A_n$. Since A_n is uniform, this yields $B \cap A_n = 0$. Now $E(A) = E(A_1 \oplus \cdots \oplus A_{n-1}) \oplus E(A_n)$, and $B \cap E(A_n) = 0$ because $A_n \leqq_e E(A_n)$; hence we obtain a monomorphism of B into $E(A_1 \oplus \cdots \oplus A_{n-1})$. Therefore $E(A_1 \oplus \cdots \oplus A_{n-1})$ has an independent family of n nonzero submodules, which contradicts the induction hypothesis. Thus A cannot have an independent family of more than n nonzero submodules.□

Definition If A is a finite-dimensional module, then in view of 3.19 there exists a nonnegative integer n such that A has an independent family of n nonzero submodules, but no independent families of more than n nonzero submodules. This integer n is called the *Goldie dimension* (or *rank*) of A, and we denote it by $\dim(A)$. For example, if R is a division ring, then $\dim(A)$ is just the vector space dimension $[A : R]$. (See also Exercises 4 and 5.) We also note that $\dim(A) = 0$ if and only if $A = 0$. As another consequence of 3.19, we see that $\dim(A) = n$ if and only if A has an essential submodule which is a direct sum of n uniform submodules. For this reason, $\dim(A)$ is sometimes called the *uniform dimension* of A. In particular, $\dim(A) = 1$ if and only if A is uniform.

Finally, to cover all situations, we set $\dim(A) = \infty$ whenever A is not finite-dimensional. Thus "finite-dimensional" now just means that $\dim(A) < \infty$, as the term suggests.

Proposition 3.20 (a) If B is a closed submodule of a module A, then $\dim(A) = \dim(B) + \dim(A/B)$.

(b) $\dim(A_1 \oplus \cdots \oplus A_n) = \dim(A_1) + \cdots + \dim(A_n)$.

(c) Let $B \leqq A$ and assume that B is finite-dimensional. Then $\dim(A) = \dim(B)$ if and only if $B \leqq_e A$.

Proof: (a) If $\dim(B) = \infty$, then obviously $\dim(A) = \infty$. If $\dim(A/B) = \infty$, then 3.13 shows that $\dim(A) = \infty$. Thus we may now assume that B and A/B are both finite-dimensional, in which case 3.13 says that A is finite-dimensional.

Let C be a relative complement for B in A, and let $C_1 \oplus \cdots \oplus C_n \leqq_e C$ with each C_i uniform. Since $C_1 \oplus \cdots \oplus C_n \oplus B \leqq_e C \oplus B \leqq_e A$, 1.4 says that $(C_1 \oplus \cdots \oplus C_n \oplus B)/B \leqq_e A/B$. Inasmuch as $(C_1 \oplus \cdots \oplus C_n \oplus B)/B$ is a direct sum of the uniform modules $(C_1 \oplus B)/B, \ldots, (C_n \oplus B)/B$, we obtain $\dim(A/B) = n$. We also have $B_1 \oplus \cdots \oplus B_k \leqq_e B$, where the B_i are uniform and $k = \dim(B)$. Now

$$C_1 \oplus \cdots \oplus C_n \oplus B_1 \oplus \cdots \oplus B_k \leqq_e A$$

and thus $\dim(A) = n + k$, as required.

(b) follows from (a) by induction.

(c) Let $B_1 \oplus \cdots \oplus B_n \leqq_e B$, where the B_i are uniform and $n = \dim(B)$. If $B \leqq_e A$, then $B_1 \oplus \cdots \oplus B_n \leqq_e A$; hence $\dim(A) = n$. Conversely, if $B \nleqq_e A$, then A has a nonzero submodule M such that $M \cap B = 0$. In this case, $\{B_1, \ldots, B_n, M\}$ is an independent family of nonzero submodules of A, whence $\dim(A) > n$.□

With the help of uniform modules, we can now proceed to show that polynomial rings over finite-dimensional rings are also finite-dimensional.

Lemma 3.21 Let f be any nonzero polynomial in $R[x]$.

(a) There exists an element $u \in R$ such that $fu \neq 0$ and the right annihilators (in R) of the nonzero coefficients of fu are all equal.

(b) If the right annihilators (in R) of the nonzero coefficients of f are all equal, then the right annihilator of f in $R[x]$ is the same as the right annihilator of its leading coefficient.

Proof: (a) Let n be the number of nonzero coefficients of f. If $n = 1$, then we can just take $u = 1$. Now let $n > 1$ and assume that (a) holds for nonzero polynomials with at most $n - 1$ nonzero coefficients.

If the right annihilators of the nonzero coefficients of f are already equal, then $u = 1$ will do. If not, then we can find nonzero coefficients r_i, r_j of f together with an element $v \in R$ such that $r_i v = 0$ but $r_j v \neq 0$. Now fv is nonzero and has at most $n - 1$ nonzero coefficients; hence there exists $w \in R$ such that $fvw \neq 0$, and the right annihilators of the nonzero coefficients of fvw are equal.

(b) Write $f = a_0 + \cdots + a_n x^n$ with $a_n \neq 0$. If $h \in R[x]$ with $a_n h = 0$, then $a_n c_j = 0$ for all coefficients c_j of h. By hypothesis, $a_i c_j = 0$ for all nonzero a_i and all c_j, from which we infer that $fh = 0$.

Now consider any $h \in R[x]$ for which $fh = 0$, and suppose that $a_n h \neq 0$. Write $h = c_0 + \cdots + c_t x^t$, and let s be the largest index for which $a_n c_s \neq 0$. Then $a_n(c_{s+1}x^{s+1} + \cdots + c_t x^t) = 0$; hence as above we obtain $f(c_{s+1}x^{s+1} + \cdots + c_t x^t) = 0$. If $h' = c_0 + \cdots + c_s x^s$, then it follows that $fh' = 0$. However, the coefficient of x^{n+s} in fh' is $a_n c_s \neq 0$, which is impossible. Therefore $a_n h = 0$.□

If A is any right ideal of R, note that the set $A[x]$ of polynomials with coefficients in A is a right ideal of $R[x]$.

Proposition 3.22 If A is a uniform right ideal of R, then $A[x]$ is a uniform right ideal of $R[x]$.

Proof: Clearly $A[x] \neq 0$. It suffices to prove that $fR[x] \leqq_e A[x]$ for any nonzero $f \in A[x]$, and we proceed by induction on $n = \deg(f)$.

If $n = 0$, then f is just a nonzero element of A, whence $fR \leqq_e A$. As R-modules, $fR[x]$ and $A[x]$ are just countably infinite direct sums of copies of fR and A; hence 1.1 says that $fR[x] \leqq_e A[x]$ as R-modules. Consequently, $fR[x] \leqq_e A[x]$ as $R[x]$-modules.

Now let $n > 0$ and assume that nonzero polynomials in $A[x]$ of degree less than n generate essential submodules of $A[x]$. If $u \in R$ is chosen as in 3.21(a), then $fuR[x] \leqq fR[x]$; hence it suffices to show that $fuR[x] \leqq_e A[x]$. In case $\deg(fu) < n$, this follows from the induction hypothesis, hence we need only consider the case when $\deg(fu) = n$. Thus, replacing f by fu, there is no loss of generality in assuming that the right annihilators of the nonzero coefficients of f are all equal.

We need $gR[x] \cap fR[x] \neq 0$ for all nonzero $g \in A[x]$, and for this we do a second induction on $k = \deg(g)$. If $k < n$ (in particular, if $k = 0$), then $gR[x] \leqq_e A[x]$ by the induction hypothesis above, whence $gR[x] \cap fR[x] \neq 0$.

Now let $k \geqq n$, and assume that $pR[x] \cap fR[x] \neq 0$ for all $p \in A[x]$ with $\deg(p) < k$. Applying 3.21 to g as we did above with f, we may assume, without loss of generality, that the right annihilators of the nonzero coefficients of g are all equal.

The respective leading terms of f and g are nonzero elements $a_n, b_k \in A$. Inasmuch as A is uniform, $a_nR \cap b_kR \neq 0$; hence we must have $a_nr = b_ks \neq 0$ for some $r,s \in R$. Observing that the right annihilators (in R) of the nonzero coefficients of fr are all equal, we see from 3.21 that fr and a_nr have the same right annihilator in $R[x]$. Likewise, gs and b_ks have the same right annihilator in $R[x]$; hence so do gs and fr.

Inasmuch as gs and $fx^{k-n}r$ both have the same leading term, namely, $b_ksx^k = a_nrx^k$, we see that the polynomial $p = gs - fx^{k-n}r$ either is zero or else has degree at most $k - 1$. If $p = 0$, then $gs \in fR[x]$, and since $b_ks \neq 0$ we obtain $gR[x] \cap fR[x] \neq 0$. If $p \neq 0$, then p is a nonzero element of $A[x]$ with $\deg(p) < k$; hence the second induction hypothesis says that $pR[x] \cap fR[x] \neq 0$. Therefore there is some $h \in R[x]$ for which ph is a nonzero element of $fR[x]$, and we observe that $gsh \in fR[x]$ as well. If $gsh = 0$, then $frh = 0$ also (because gs and fr have the same right annihilator in $R[x]$). But then $ph = gsh - x^{k-n}frh = 0$, which is a contradiction. Thus $gsh \neq 0$, from which we conclude that $gR[x] \cap fR[x] \neq 0$.

This completes the induction on k; hence we have $gR[x] \cap fR[x] \neq 0$ for all nonzero $g \in A[x]$. Consequently $fR[x] \leqq_e A[x]$, which completes the induction on n. Therefore $A[x]$ is indeed uniform.□

Theorem 3.23 Let X be any collection of indeterminates. If R_R is finite-dimensional, then so is $R[X]_{R[X]}$, and $\dim(R_R) = \dim(R[X]_{R[X]})$.

Proof: We first claim that if A is any uniform right ideal of R, then $A[X]$ is a uniform right ideal of $R[X]$. Obviously $A[X] \neq 0$. Given any nonzero polynomials $f,g \in A[X]$, we may choose a finite number of indeterminates $x_1, \ldots, x_n \in X$ such that $f,g \in A[x_1, \ldots, x_n]$. By induction on 3.22, $A[x_1, \ldots, x_n]$ is a uniform right ideal of $R[x_1, \ldots, x_n]$; hence

$$fR[x_1, \ldots, x_n] \cap gR[x_1, \ldots, x_n] \neq 0$$

and thus $fR[X] \cap gR[X] \neq 0$. Therefore $A[X]$ is uniform.

Now $A_1 \oplus \cdots \oplus A_n \leqq_e R_R$ for some uniform right ideals A_i, where $n = \dim(R_R)$. Since $R[X]$ is a free right R-module with a basis consisting of all monomials $x_1 x_2 \cdots x_n$, we see from 1.1 that $(A_1 \oplus \cdots \oplus A_n)[X] \leqq_e R[X]$ as right R-modules. Consequently, $A_1[X] \oplus \cdots \oplus A_n[X] \leqq_e R[X]$ as right $R[X]$-modules. Inasmuch as each $A_i[X]$ is a uniform right ideal of $R[X]$, we conclude that $R[X]_{R[X]}$ is finite-dimensional with dimension n.□

Returning to nonsingular modules and rings, we have the following results, analogous to 3.16 and 3.17.

Proposition 3.24 Let $Z_r(R) = 0$. For any right R-module A, the following conditions are equivalent:

(a) $S^\circ A$ is a simple right $S^\circ R$-module.

(b) $A/Z(A)$ is uniform.

(c) $L^*(A) = \{A,Z(A)\}$ and $Z(A) < A$.

Proof: (a) ⇒ (b): If B is a nonzero submodule of $A/Z(A)$, then $S^\circ B$ is a nonzero $S^\circ R$-submodule of $S^\circ A$, whence $S^\circ B = S^\circ A$. Then $B \leqq_e S^\circ A$, and consequently $B \leqq_e A/Z(A)$. Therefore $A/Z(A)$ is uniform.

(b) ⇒ (c): Since $A/Z(A) \neq 0$, $Z(A) < A$. Inasmuch as $Z(A/Z(A)) = 0$ by 1.23, we have $0, A/Z(A) \in L^*(A/Z(A))$. Any other submodule of $A/Z(A)$ is a proper essential submodule, and hence cannot be $\mathcal{S}$-closed. Therefore $L^*(A/Z(A)) = \{0,A/Z(A)\}$; hence it follows from 3.15 that $L^*(A) = \{Z(A),A\}$.

(c) ⇒ (a): According to 3.15, $L^*(S^\circ A) = \{0,S^\circ A\}$. Given any nonzero $x \in S^\circ A$, we have $x(S^\circ R) \in L^*(S^\circ A)$ by 2.9, whence $x(S^\circ R) = S^\circ A$. Therefore $(S^\circ A)_{S^\circ R}$ is simple.□

Corollary 3.25 If $Z_r(R) = 0$, then the following conditions are equivalent:

(a) $S^\circ R$ is a division ring.

(b) R_R is uniform.

(c) R is an integral domain and R_R is finite-dimensional.

Proof: (a) ⇔ (b) by 3.24.

(a) ⇒ (c): Since R is a subring of a division ring, it must be an integral domain. Also, R_R is uniform by (b) and thus is finite-dimensional.

(c) ⇒ (b): According to 3.18, R must have a uniform right ideal I. Inasmuch as R is a domain, R_R is isomorphic to a nonzero submodule of I and consequently is uniform.□

More generally, we need information about the simple submodules of $S°A$, which we relate to $L^*(A)$ in terms of the following concepts.

Definition Let P be a partially ordered set with a smallest element 0. An *atom* in P is an element x which is minimal with respect to the property $x > 0$. For example, if P is the family of all subsets of some set X, then the empty set is the smallest element of P, and the atoms in P are just the singletons. On the other hand, the unit interval [0,1] has no atoms.

We say that P is *atomic* provided every nonzero element of P lies above an atom of P. For example, the family of all subsets of a set is always atomic.

Lemma 3.26 Let $Z_r(R) = 0$. For any right R-module A, the atoms in $L^*(S°A)$ are exactly the simple $S°R$-submodules of $S°A$.

Proof: If K is an atom of $L^*(S°A)$, then K is an $S°R$-submodule of $S°A$ by 2.9. Given any nonzero $x \in K$, we have $x(S°R) \in L^*(S°A)$ by 2.9, and thus $x(S°R) = K$. Therefore $K_{S°R}$ is simple.

Conversely, let K be a simple $S°R$-submodule of $S°A$, and note by 2.9 that $K \in L^*(S°A)$. If $M \in L^*(S°A)$ with $M \leqq K$, then M is an $S°R$-submodule of K by 2.9; hence either $M = 0$ or $M = K$. Therefore K is an atom of $L^*(S°A)$. □

Theorem 3.27 Let $Z_r(R) = 0$. For any right R-module A, the following conditions are equivalent:

(a) $(S°A)_{S°R}$ has essential socle.

(b) $L^*(A)$ is atomic.

(c) Every nonzero submodule of $A/Z(A)$ contains a uniform submodule.

Proof: If $B = A/Z(A)$, then $S°B = S°A$, $L^*(B) \cong L^*(A)$ by 3.15, and $Z(B) = 0$ by 1.23. Thus, replacing A by B, there is no loss of generality in assuming that $Z(A) = 0$.

(a) ⇒ (b): In view of 3.15, it suffices to prove that $L^*(S°A)$ is atomic. But any nonzero $M \in L^*(S°A)$ is an $S°R$-submodule of $S°A$ by 2.9; hence $M \cap \operatorname{soc}((S°A)_{S°R}) \neq 0$. Then M contains a simple $S°R$-submodule K, which by 3.26 is an atom of $L^*(S°A)$.

(b) ⇒ (c): Given any nonzero $B \leqq A$, let M denote the $\mathscr{S}$-closure of B in A, and note by 2.3 that $B \leqq_e M$. According to (b), M must contain an atom K of $L^*(A)$. Since the $\mathscr{S}$-closure in A of any nonzero submodule of K is clearly K, we infer from 2.3 that all nonzero submodules of K are essential in K, that is, K is uniform. Now $K \cap B \neq 0$ because $B \leqq_e M$; hence $K \cap B$ is a uniform submodule of B.

(c) ⇒ (a): If M is a nonzero $S°R$-submodule of $S°A$, then $M \cap A$ is a nonzero R-submodule of A and thus contains a uniform submodule K. Inasmuch as K is essentially finitely generated, 2.10 says that $S°K = K(S°R)$, whence $S°K \leqq M$. According to 3.24, $(S°K)_{S°R}$ is simple, hence we conclude that $M \cap \text{soc}((S°A)_{S°R}) \neq 0$.□

In particular, 3.27 gives necessary and sufficient conditions under which the ring $S°R$ has essential socle. This, in turn, we show to be equivalent to decomposing $S°R$ into a direct product of rings of the following kind.

Definition A *right full linear ring* is the ring of all linear transformations (written on the *left*) of a *right* vector space over a division ring. In particular, every simple artinian ring is isomorphic to a right full linear ring.

Theorem 3.28 *R is isomorphic to a direct product of right full linear rings if and only if R is a regular, right self-injective ring and* $\text{soc}(R_R) \leqq_e R_R$.

Proof: First assume that $R = \Pi R_\gamma$, where R_γ is the ring of all linear transformations on a right vector space V_γ over a division ring D_γ. By 2.23, each R_γ is a regular, right self-injective ring, whence the same holds for R.

Given any nonzero $f \in R$, we have $f_\lambda \neq 0$ for some λ; hence $f_\lambda v \neq 0$ for some $v \in V_\lambda$. Choose an idempotent $e_\lambda \in R_\lambda$ which projects onto the one-dimensional subspace vD_λ, and note that $f_\lambda e_\lambda \neq 0$ and that $e_\lambda R_\lambda e_\lambda \cong D_\lambda$. We claim that $e_\lambda R_\lambda$ is a simple right R_λ-module. Given any nonzero $g \in e_\lambda R_\lambda$, we have $(gR_\lambda)^2 = gR_\lambda \neq 0$ because R_λ is regular, from which we infer that $e_\lambda g R_\lambda e_\lambda \neq 0$. Inasmuch as $e_\lambda R_\lambda e_\lambda$ is a division ring, we must have

$$(e_\lambda g R_\lambda e_\lambda)(e_\lambda R_\lambda e_\lambda) = e_\lambda R_\lambda e_\lambda$$

and consequently $gR_\lambda = e_\lambda g R_\lambda = e_\lambda R_\lambda$. Therefore $e_\lambda R_\lambda$ is simple, as claimed. Setting $e_\gamma = 0$ for all $\gamma \neq \lambda$, we obtain an element $e \in R$ for which $(eR)_R$ is simple; hence $e \in \text{soc}(R_R)$. Since $fe \neq 0$ and $\text{soc}(R_R)$ is a two-sided ideal, it follows that $fR \cap \text{soc}(R_R) \neq 0$. Therefore $\text{soc}(R_R) \leqq_e R_R$.

Conversely, assume that R is regular and right self-injective with essential socle. If X denotes the set of those idempotents $e \in R$ for which eRe is a division ring, then we claim that any nonzero two-sided ideal I of R must contain an element of X. Now $I \cap \text{soc}(R_R) \neq 0$ because $\text{soc}(R_R) \leqq_e R_R$; hence I contains a simple right R-module A. Since A is a principal right ideal in the regular ring R, we must have $A = eR$ for some idempotent $e \in R$. Then $eRe \cong \text{End}_R(A)$, which is a division ring by Schur's Lemma, whence $e \in X \cap I$, as desired.

Let $\mathfrak{A} = \{ReR \mid e \in X\}$, and let $\{Re_\gamma R\}$ be a maximal independent subfamily of $\mathfrak{A}$. If K denotes the left annihilator of $\oplus Re_\gamma R$, then

$$K \cap (\oplus Re_\gamma R) = [K \cap (\oplus Re_\gamma R)]^2 \leqq K(\oplus Re_\gamma R) = 0$$

Now K is a two-sided ideal of R; so if $K \neq 0$, we must have $ReR \leqq K$ for some $e \in X$. But then $\{Re_\gamma R, ReR\}$ is independent, which contradicts the maximality of $\{Re_\gamma R\}$. Therefore $K = 0$. Note also that

$$(Re_\lambda R)(Re_\gamma R) \leqq (Re_\lambda R) \cap (Re_\gamma R) = 0$$

whenever $\lambda \neq \gamma$.

For each γ, let D_γ be the division ring $e_\gamma R_\gamma e_\gamma$, V_γ the vector space Re_γ, and R_γ the ring of all linear transformations on V_γ. There is an obvious ring map $\phi : R \to \Pi R_\gamma$, where $(\phi x)_\gamma$ is the endomorphism of V_γ given by left multiplication by x. Now $\ker \phi$ is the left annihilator of $\oplus Re_\gamma$, which is the same as the left annihilator of $\oplus Re_\gamma R$. We have just shown that this annihilator is zero, whence ϕ must be injective.

Now ϕR is isomorphic to R and so is right self-injective; hence $(\phi R)_{\phi R}$ has no proper essential extensions. Thus to show that ϕ is surjective, it suffices to show that $(\phi R)_{\phi R} \leqq_e (\Pi R_\gamma)_{\phi R}$.

Given any nonzero $f \in \Pi R_\gamma$, we have $f_\lambda \neq 0$ for some λ, and then $f_\lambda v \neq 0$ for some $v \in V_\lambda$. Now $vV_\gamma \leqq (Re_\lambda R)(Re_\gamma R) = 0$ for all $\gamma \neq \lambda$, and since $f_\lambda v \in V_\lambda$ we likewise have $(f_\lambda v)V_\gamma = 0$ for all $\gamma \neq \lambda$. Therefore $(\phi v)_\gamma = [\phi(f_\lambda v)]_\gamma = 0$ for all $\gamma \neq \lambda$. For any $r \in R$, we compute that

$$\begin{aligned}[\phi(f_\lambda v)]_\lambda(re_\lambda) &= (f_\lambda v)(re_\lambda) = [f_\lambda(ve_\lambda)](re_\lambda) \\ &= (f_\lambda v)(e_\lambda re_\lambda) = f_\lambda(ve_\lambda re_\lambda) \\ &= f_\lambda(vre_\lambda) = f_\lambda(\phi v)_\lambda(re_\lambda)\end{aligned}$$

Thus $[\phi(f_\lambda v)]_\lambda = f_\lambda(\phi v)_\lambda$, from which we conclude that $\phi(f_\lambda v) = f(\phi v)$. Since $f_\lambda v \neq 0$, $f(\phi v)$ is thus a nonzero element of ϕR.

Therefore $(\phi R)_{\phi R} \leqq_e (\Pi R_\gamma)_{\phi R}$, and consequently ϕ is surjective. Thus ϕ is a ring isomorphism of R onto ΠR_γ. □

Theorem 3.29 If $Z_r(R) = 0$, then the following conditions are equivalent:

(a) $S^\circ R$ is isomorphic to a direct product of right full linear rings.
(b) $L^*(R_R)$ is atomic.
(c) Every nonzero right ideal of R contains a uniform right ideal.

Proof: Inasmuch as $S^\circ R$ is regular and right self-injective, 3.28 shows that (a) is equivalent to the condition $\operatorname{soc}((S^\circ R)_{S^\circ R}) \leqq_e (S^\circ R)_{S^\circ R}$. By 3.27, this is equivalent to (b) and to (c). □

It is clear that 3.17(c) implies 3.29(a). The converse of course is false; for example, the ring R in 3.11 satisfies 3.29 because $S^\circ R$ is a direct product of fields, while 3.17 fails because $S^\circ R$ is not semisimple. We also note that the conditions of 3.29 do not hold universally, since there exist right nonsingular rings with no uniform right ideals (Exercise 6).

Exercises

1. Let R be a commutative integral domain with quotient field K. Show that a torsion-free R-module A is uniform if and only if $A \otimes_R K$ is a one-dimensional vector space over K.
2. Given an integer $n > 1$, show that $\mathbf{Z}/n\mathbf{Z}$ is uniform if and only if n is a power of a prime.
3. Find an example of modules $B \leqq A$ such that A is uniform but A/B is not.
4. Let R be a commutative integral domain with quotient field K. For any torsion-free R-module A, show that $\dim(A) = [A \otimes_R K : K]$.
5. Given an integer $n > 1$, show that $\dim(\mathbf{Z}/n\mathbf{Z})$ is the number of prime divisors of n.
6. Let $F_1, F_2, \ldots$ be fields, $R = (\Pi F_n)/(\oplus F_n)$. Show that $Z_r(R) = 0$ and that R has no uniform ideals.
7. Show that a nonzero module A is uniform if and only if $E(A)$ is indecomposable.
8. Show that a module A is uniform if and only if A has exactly two closed submodules.
9. Show that a right ideal in a regular ring is uniform if and only if it is simple.
10. Prove that a nonzero quasi-injective module is uniform if and only if it is indecomposable.
11. Prove 3.18 directly from the definitions, without using 3.14.
12. Prove that any finite-dimensional quasi-injective module is a direct sum of uniform quasi-injective modules.
13. If $Z_r(R) = 0$, prove that every uniform right R-module is either singular or nonsingular.
14. If Q is a right quotient ring of R, prove that $\dim(Q_Q) = \dim(R_R)$.
15. Let $n > 0$. If T is the ring of all $n \times n$ matrices over R, prove that $\dim(T_T) = n \cdot \dim(R_R)$.
16. Do Exercise 15 for the ring of all lower triangular $n \times n$ matrices over R.
17. Suppose that $A_1 \oplus \cdots \oplus A_n \leqq_e A$ with each A_i uniform. If $Z(A) = 0$, prove that $L^*(A)$ has a composition series of length n.
18. If $Z(A) = 0$, use the Jordan-Hölder-Dedekind Theorem in combination with Exercise 17 to prove 3.19(b). (*Hint*: Exercise 2.A.11.)
19. If $Z_r(R) = 0$ and R_R is finite-dimensional, use the Hilbert Basis Theorem to prove that $R[x]_{R[x]}$ is finite-dimensional.
20. Let $Z_r(R) = 0$; then $Z_r(R[x]) = 0$ by Exercise 1.D.13. If $L^*(R_R)$ is atomic, prove that $L^*(R[x]_{R[x]})$ is atomic also.
21. Let $Z_r(R) = 0$ and assume that $L^*(R_R)$ is atomic. If every closed one-sided ideal of R is an annihilator, prove that $S^\circ R$ is isomorphic to a direct product of full matrix rings over division rings.
22. Two uniform modules A,B are *equivalent* if $E(A) \cong E(B)$. If R_R is finite-dimensional, prove that the number of equivalence classes of uniform right ideals of R is at most $\dim(R_R)$.
23. Let $Z_r(R) = 0$, $R \neq 0$. If R_R is finite-dimensional, prove that $S^\circ R$ is a direct product of n simple artinian rings, where n is the number of equivalence classes of uniform right ideals of R.

D. Goldie Rings

For a commutative integral domain R, the construction of the quotient field of R is relatively straightforward. If R is allowed to be noncommutative, however, there are obstacles to imitating this procedure. The goal is clear enough: We would like to embed R in a division ring in such a way that every element of D can be written as a "fraction" with both numerator and denominator coming from R. First, we must agree on how to write these fractions, since ab^{-1} could easily be different from $b^{-1}a$. Then the major difficulty comes in trying to write down definitions for addition and multiplication. For example, what should a product $(ab^{-1})(cd^{-1})$ look like if $b^{-1}c \neq cb^{-1}$? Fortunately, there is a simple condition on R which is sufficient to overcome these difficulties.

Definition A *right Ore domain* is an integral domain R such that any two nonzero elements of R have a nonzero common right multiple, that is, $aR \cap bR \neq 0$ for all nonzero $a,b \in R$. Thus an integral domain R is a right Ore domain if and only if R_R is uniform. For example, every commutative integral domain is a right Ore domain. *Left Ore domains* are defined similarly, but not all right Ore domains are left Ore domains (Exercise 1).

The classical procedure for constructing a quotient division ring D from a right Ore domain R runs as follows. First set $X = R \times (R - \{0\})$, and define an equivalence relation "$\sim$" on X, where $(a,b) \sim (c,d)$ if and only if there exist nonzero elements $r,s \in R$ such that $ar = cs$ and $br = ds$. Let $[a,b]$ denote the equivalence class of (a,b), and let D denote the collection of these equivalence classes. To add $[a,b]$ and $[c,d]$, choose nonzero elements $r,s \in R$ such that $br = ds$, and define $[a,b] + [c,d] = [ar + cs, br]$. To multiply $[a,b]$ and $[c,d]$, choose elements $r,s \in R$ such that $s \neq 0$ and $br = cs$, and define $[a,b][c,d] = [ar,ds]$. Finally, there is the tedious task of checking that all this is well-defined, that D satisfies the ring axioms, and that D is a division ring. The ring R is identified with the subring $\{[r,1] \mid r \in R\}$ of D, and then every element of D has the form ab^{-1} for suitable $a,b \in R$.

Fortunately, we can avoid this entire procedure. According to 2.11, a quotient division ring for R, if one exists, must equal $S^{\circ}R$. Thus we need only show that $S^{\circ}R$ is a division ring consisting of fractions of elements of R.

Theorem 3.30 If R is an integral domain, then the following conditions are equivalent:

(a) There is a division ring D, containing R as a subring, such that $D = \{ab^{-1} \mid a,b \in R, b \neq 0\}$.

(b) R is a right Ore domain.

(c) R_R is finite-dimensional.

Proof: Since $Z_r(R) = 0$ by 1.27, we have (b) $\Leftrightarrow$ (c) by 3.25.

(a) $\Rightarrow$ (b): Consider any nonzero elements $x,y \in R$. Since D is a division ring, there exists $x^{-1}y \in D$, and then (a) says that $x^{-1}y = ab^{-1}$ for some $a,b \in R$, $b \neq 0$. Multiplying this out, we obtain $xa = yb$, and consequently $xR \cap yR \neq 0$.

(b) $\Rightarrow$ (a): According to 3.25, $S°R$ is a division ring. Given any nonzero $x \in S°R$, we have $xR \cap R \neq 0$ and thus $xb \in R$ for some nonzero $b \in R$, whence $x = (xb)b^{-1}$. Therefore $S°R = \{ab^{-1} \mid a,b \in R,\ b \neq 0\}$.□

Note that 3.30(a) is *not* left-right symmetric; for a left Ore domain, the fractions in the quotient division ring all have the form $b^{-1}a$. Unless R is a right Ore domain as well, some of the fractions $b^{-1}a$ will not be expressible in the form cd^{-1} (see Exercise 2).

For a ring R which is not an integral domain, we must be careful about which elements of R may be expected to be invertible in a ring containing R. For example, if $xy = 0$ in R with $x,y \neq 0$, then neither x nor y can be invertible. Thus the most we can look for is a ring in which all non-zero-divisors of R have inverses.

Definition A *regular element* (or a *non-zero-divisor*) in R is an element x such that $rx \neq 0$ and $xr \neq 0$ for all nonzero $r \in R$.

A *classical right quotient ring* for R is a ring Q which contains R as a subring in such a way that every regular element of R is invertible in Q and

$$Q = \{ab^{-1} \mid a,b \in R,\ b \text{ regular}\}$$

In order not to conflict with previous terminology, we must show that such a ring Q is a right quotient ring of R. Thus consider any $x,y \in Q$ with $x \neq 0$, and write $y = ab^{-1}$ with $a,b \in R$, b regular. Then $yb = a \in R$, and $xb \neq 0$ because b is invertible in Q. Therefore $R_R \leq_r Q_R$; hence Q is indeed a right quotient ring of R. On the other hand, Q need not be a left quotient ring of R (see Exercise 3).

For example, the quotient field of a commutative integral domain R is a classical right quotient ring of R. Likewise, if R is a right Ore domain, it follows from the proof of 3.30 that $S°R$ is a classical right quotient ring of R. In fact, an integral domain has a classical right quotient ring if and only if it is right Ore (Exercise 4). Since there exist integral domains which are neither right Ore nor left Ore (Exercise 5), we thus get examples of rings which have no classical right quotient rings and no classical left quotient rings. For a right Ore domain R, we have noticed that R has a classical right quotient ring $S°R$, which happens to be the maximal right quotient ring of R. In general, however, even if a ring R has a classical right quotient ring Q, Q need not be isomorphic to the maximal right quotient ring of R (Exercise 6).

A general criterion for the existence of a classical right quotient ring is given in Exercise 8. We concentrate now on a problem more related to the Ore domain situation; namely, when does R have a classical right quotient ring which is a semisimple ring? As in 3.30, our work is made substantially easier because a semisimple classical right quotient ring, if one exists, must coincide with $S^\circ R$.

Definition A *right Goldie ring* is a ring R such that R_R is finite-dimensional and such that the right annihilator ideals in R satisfy the ACC. For example, every right noetherian ring is a right Goldie ring.

Any integral domain R automatically has ACC on right annihilators, since 0 and R are the only right annihilator ideals in R. Thus R is right Goldie if and only if R_R is finite-dimensional, which by 3.30 happens if and only if R is right Ore. Consequently, the integral domain of Exercise 1 is an example of a right Goldie ring which is not left Goldie.

Definition Given $x \in R$, we write $r(x)$ for the right annihilator $\{r \in R \mid xr = 0\}$. Similarly, we write $r(X)$ for the right annihilator of a subset X of R.

Proposition 3.31 If R has ACC on right annihilators, then $Z_r(R)$ is nilpotent.

Proof: Let $J = Z_r(R)$, and suppose that J is not nilpotent. We claim that as a consequence, $r(J^k) < r(J^{k+1})$ for every positive integer k.

By assumption, $J^{k+1} \neq 0$; hence the set $X = \{x \in J \mid J^k x \neq 0\}$ is nonempty. Inasmuch as R has ACC on right annihilators, we may pick an element $x \in X$ such that $r(x)$ is maximal among the right annihilators of elements of X. Given any $a \in J$, we have $r(a) \in \mathscr{S}(R)$ by definition of J, whence $r(a) \cap xR \neq 0$. Thus there exists an element $s \in R$ for which $xs \neq 0$ and $xs \in r(a)$, that is, $xs \neq 0$ but $axs = 0$. Then $r(x) < r(ax)$; hence by maximality of $r(x)$ we must have $ax \notin X$. Since $ax \in J$, this means that $J^k ax = 0$. Inasmuch as this holds for any $a \in J$, we obtain $J^{k+1}x = 0$, and thus $x \in r(J^{k+1})$. On the other hand, $J^k x \neq 0$ because $x \in X$, hence we conclude that $r(J^k) < r(J^{k+1})$.

Therefore $r(J) < r(J^2) < \cdots$, which contradicts the hypothesis that R has ACC on right annihilators. □

Corollary 3.32 Let R be semiprime. Then R is a right Goldie ring if and only if $Z_r(R) = 0$, and R_R is finite-dimensional.

Proof: If R is right Goldie, then R_R is finite-dimensional by definition. According to 3.31, $Z_r(R)$ is a nilpotent ideal of R, which because of semiprimeness implies that $Z_r(R) = 0$.

Conversely, assume that $Z_r(R) = 0$ and that R_R is finite-dimensional. We claim that any right annihilator ideal $r(X)$ must belong to $L^*(R_R)$. For if

$a \in R$ and $I \in \mathcal{S}(R)$ with $aI \leqq r(X)$, then $XaI = 0$ and consequently $Xa \subseteq Z_r(R) = 0$, whence $a \in r(X)$. Therefore $r(X) \in L^*(R_R)$. Inasmuch as $L^*(R_R)$ has ACC by 3.16, we conclude that R must have ACC on right annihilators.□

Lemma 3.33 Let $Z_r(R) = 0$, and assume that R_R is finite-dimensional. For any $x \in R$, the following conditions are equivalent:

(a) x is regular in R.
(b) $r(x) = 0$.
(c) $xR \in \mathcal{S}(R)$.
(d) x is invertible in $S^\circ R$.

Proof: (d) ⇒ (a) ⇒ (b) is clear.

(b) ⇒ (c): For $n = 1, 2, \ldots$, let K_n be the $\mathcal{S}$-closure of $x^n R$ in R_R, whence $K_1 \geqq K_2 \geqq \cdots$ in $L^*(R_R)$. According to 3.16, $L^*(R_R)$ has DCC; hence we obtain $K_n = K_{n+1}$ for some n. Then by 2.3 we have $x^{n+1}R \leqq_e K_{n+1} = K_n$, from which we infer that $x^{n+1}R \leqq_e x^n R$. Inasmuch as $r(x) = 0$, we must have $r(x^n) = 0$ as well; hence left multiplication by x^n provides an isomorphism of R_R onto $x^n R$ which carries xR onto $x^{n+1}R$. Applying the inverse isomorphism, we conclude that $xR \leqq_e R_R$.

(c) ⇒ (d): Setting $Q = S^\circ R$, we have $xRQ = Q$ by 3.17(e), whence $xQ = Q$. Therefore x is right invertible in Q. As a consequence, we see that ${}_QQ \cong Qx$, hence $\dim({}_QQ) = \dim({}_Q(Qx))$. Now Q is a semisimple ring by 3.17 and so is left noetherian, whence Qx must be a finite-dimensional left Q-module. According to 3.20, we must have $Qx \leqq_e {}_QQ$. Inasmuch as ${}_QQ$ is semisimple, we obtain $Qx = Q$, and thus x is also left invertible in Q.□

Theorem 3.34 Let R be a semiprime right Goldie ring, and let I be any right ideal of R. Then $I \in \mathcal{S}(R)$ if and only if I contains a regular element.

Proof: If there is a regular element $x \in I$, then $xR \in \mathcal{S}(R)$ by 3.33, whence $I \in \mathcal{S}(R)$.

Conversely, assume that $I \in \mathcal{S}(R)$. As noted in 3.32, all right annihilator ideals in R belong to $L^*(R_R)$; hence in view of 3.16 we see that R has DCC on right annihilators. Thus there is an element $x \in I$ such that $r(x)$ is minimal among the right annihilators of elements of I. We claim that x is regular, and by 3.33 it suffices to prove that $xR \in \mathcal{S}(R)$. Since $I \in \mathcal{S}(R)$, this follows if we show that $xR \leqq_e I$.

Thus consider any $M \leqq I$ for which $M \cap xR = 0$. Given any $m \in M$, we have $mR \cap xR = 0$, from which we infer that $r(m + x) = r(m) \cap r(x) \leqq r(x)$. Since $m + x \in I$, it follows from the minimality of $r(x)$ that $r(m + x) = r(x)$. Therefore $r(x) = r(m) \cap r(x) \leqq r(m)$, whence $m[r(x)] = 0$. Inasmuch as this holds for all $m \in M$, we have $Mr(x) = 0$, and consequently $[r(x)M]^2 = 0$. Since R is semiprime, it follows that $r(x)M = 0$.

We now claim that $\{M,xM,x^2M,\ldots\}$ is an independent sequence of right ideals of R. Obviously $\{M\}$ is independent. Assume that $\{M,xM,\ldots,x^nM\}$ is independent for some $n \geqq 0$. Inasmuch as $r(x)$ is a right ideal of R, we see that

$$[(M \oplus xM \oplus \cdots \oplus x^nM) \cap r(x)]^2 \leqq r(x)M + r(x)xM + \cdots + r(x)x^nM$$
$$\leqq r(x)M = 0$$

hence by semiprimeness we obtain $(M \oplus xM \oplus \cdots \oplus x^nM) \cap r(x) = 0$. Therefore left multiplication by x provides an isomorphism of $M \oplus xM \oplus \cdots \oplus x^nM$ onto $xM + \cdots + x^{n+1}M$, whence $\{xM,\ldots,x^{n+1}M\}$ is independent. Inasmuch as $M \cap (xM \oplus \cdots \oplus x^{n+1}M) \leqq M \cap xR = 0$, we conclude that $\{M,xM,\ldots,x^{n+1}M\}$ is independent. Thus the induction works, and consequently $\{M,xM,x^2M,\ldots\}$ must be independent.

Since R_R is finite-dimensional, it follows that $x^nM = 0$ for some $n \geqq 0$. If $n > 0$, then $x^{n-1}M \leqq r(x)$ and so $x^{n-1}M^2 \leqq r(x)M = 0$. Continuing in this fashion, we find that $M^{n+1} = 0$; hence we conclude from semiprimeness that $M = 0$.

Therefore $xR \leqq_e I \leqq_e R_R$, whence $xR \in \mathscr{S}(R)$. By 3.33, x is regular. □

Theorem 3.35 The following conditions are equivalent:

(a) R has a classical right quotient ring which is semisimple.

(b) R is a semiprime right Goldie ring.

(c) $Z_r(R) = 0$ and $\mathscr{S}(R)$ has a cofinal subset of principal right ideals.

Proof: (c) ⇒ (b): According to 3.17, R_R is finite-dimensional. In view of 3.32, all that remains is to show that R is semiprime. If not, then R has a nonzero two-sided ideal I such that $I^2 = 0$. Choosing a right ideal J for which $I_R \oplus J \leqq_e R_R$, we note that $JI \leqq J \cap I = 0$, whence $(I \oplus J)I = 0$. According to (c), $I \oplus J$ contains a member of $\mathscr{S}(R)$ of the form xR, and we see by 3.33 that x must be regular. But $xI \leqq (I \oplus J)I = 0$ and thus $I = 0$, which is a contradiction. Therefore R is semiprime.

(b) ⇒ (a): According to 3.32, $Z_r(R) = 0$. Since R_R is finite-dimensional, 3.33 shows that all regular elements of R are invertible in $S°R$. Given any $x \in S°R$, we must have $xI \leqq R$ for some $I \in \mathscr{S}(R)$. Now I contains a regular element b by 3.34 and thus $x = ab^{-1}$, where $a = xb \in R$. Therefore $S°R$ is a classical right quotient ring of R. Also, $S°R$ is a semisimple ring, by 3.17.

(a) ⇒ (c): Let Q be a semisimple classical right quotient ring of R. Inasmuch as Q is regular and right self-injective, 2.11 says that $Z_r(R) = 0$ and $S°R = Q$. Since Q is semisimple, 3.17 shows that R_R is finite-dimensional.

We claim that there exists a common denominator for any finite set of elements $x_1,\ldots,x_n \in Q$, that is, a regular element $c \in R$ such that all $x_i \in Rc^{-1}$. If $n = 1$, there is nothing to prove. Now let $n > 1$ and assume that we have $x_i = b_i c^{-1}$ for $i = 1, \ldots, n-1$, where $b_1, \ldots, b_{n-1}, c \in R$ and c is regular. We also have $x_n = tu^{-1}$ for some $t,u \in R$, u regular. Now $c^{-1}u = de^{-1}$ for some $d,e \in R$ with e regular, and consequently $cd = ue$. We have $x_i = (b_i d)(ue)^{-1}$ for $i = 1, \ldots, n-1$ and $x_n = (te)(ue)^{-1}$; hence ue is a common denominator for $x_1, \ldots, x_n$, and the induction works.

Given any $I \in \mathscr{S}(R)$, we have $IQ = Q$ by 3.17, whence $a_1x_1 + \cdots + a_nx_n = 1$ for some $a_i \in I$, $x_i \in Q$. As we have just shown, there exist elements $b_1, \ldots, b_n, c \in R$, with c regular, such that $x_i = b_ic^{-1}$ for all i. Consequently $c = a_1b_1 + \cdots + a_nb_n \in I$. Then $cR \leqq I$, and $cR \in \mathscr{S}(R)$ by 3.33. Therefore every member of $\mathscr{S}(R)$ contains a member of $\mathscr{S}(R)$ which is principal.□

The equivalence of (a) and (b) in 3.35 is known as *Goldie's Theorem*, or *Goldie's Second Theorem*. *Goldie's First Theorem* is the following special case.

Corollary 3.36 *R has a classical right quotient ring which is simple artinian if and only if R is a prime right Goldie ring.*

Proof: If R is prime right Goldie, then by 3.35 R at least has a classical right quotient ring Q which is semisimple. Now Q is a finite direct product of simple artinian rings; hence if Q is not simple artinian, there must be a ring decomposition $Q = Q_1 \times Q_2$ such that each $Q_i \neq 0$. Inasmuch as Q is a right quotient ring of R, $R_R \leqq_e Q_R$, and consequently each $Q_i \cap R \neq 0$. Then $Q_1 \cap R$ and $Q_2 \cap R$ are nonzero two-sided ideals of R satisfying $(Q_1 \cap R)(Q_2 \cap R) = 0$, which contradicts the hypothesis that R is prime. Therefore Q must be simple artinian.

Conversely, assume that R has a classical right quotient ring Q which is simple artinian. According to 3.35, R is at least a semiprime right Goldie ring. If R is not prime, then we have $AB = 0$ for some nonzero two-sided ideals A,B of R. Now QAQ is a nonzero two-sided ideal in the simple ring Q; hence $QAQ = Q$. Consequently, $x_1a_1y_1 + \cdots + x_na_ny_n = 1$ for some $x_i, y_i \in Q$, $a_i \in A$. Inasmuch as $R_R \leqq_e Q_R$, there exists some $I \in \mathscr{S}(R)$ such that $y_iI \leqq R$ for all i. According to 3.34, there must be a regular element $c \in I$, and since $y_ic \in R$ for all i we see that $c \in QA$. Thus $cB \leqq QAB = 0$; hence it follows from the regularity of c that $B = 0$, which is false. Therefore R must be prime.□

Unlike the situation for maximal quotient rings, classical right and left quotient rings of R always coincide, provided both exist (Exercise 9). For semiprime right and left Goldie rings, the maximal right and left quotient rings coincide also, as we now show.

Theorem 3.37 *Let R be a semiprime right and left Goldie ring. Then there* exists a semisimple ring Q which is a classical right and left quotient ring of R as well as a maximal right and left quotient ring of R.

Proof: According to 3.35, R has a classical right quotient ring Q which is semisimple. Inasmuch as Q_Q is injective, it has no proper rational extensions; hence Q has no proper right quotient rings. Consequently, 2.30 says that Q is a maximal right quotient ring of R.

Using 3.35 on the left, we see that R also has a classical left quotient ring P.

Any element $x \in Q$ has the form ab^{-1} for suitable $a,b \in R$, b regular. Inasmuch as b is also invertible in P, we obtain $ab^{-1} = c^{-1}d$ *in* P for some $c,d \in R$, c regular. As a consequence, $ca = db$ *in* R, and thus $x = ab^{-1} = c^{-1}d$ *in* Q. Therefore

$$Q = \{c^{-1}d \mid c,d \in R,\ c \text{ regular}\}$$

whence Q is a classical left quotient ring of R. Since Q is semisimple, we conclude as above that Q is also a maximal left quotient ring of R.□

Exercises

1. Let F be a field such that there exists an isomorphism λ of F onto a proper subfield of F. Let R be the abelian group consisting of all polynomials in x with coefficients from F, with *coefficients written on the right*. Define a multiplication in R by using the rule $ax^n = x^n(\lambda^n a)$ for all $a \in F$ and all n. Prove that R is a principal right ideal domain, and that R is right Ore but not left Ore.
2. Let R be an integral domain. Prove that R is both right and left Ore if and only if there exists a division ring D, containing R, such that $D = \{ab^{-1} \mid a,b \in R,\ b \neq 0\}$ and also $D = \{b^{-1}a \mid a,b \in R,\ b \neq 0\}$.
3. Let R be a right Ore domain which is not left Ore. Show that $S^\circ R$ is a classical right quotient ring of R, but that $S^\circ R$ is not a left quotient ring of R.
4. Show that an integral domain has a classical right quotient ring if and only if it is right Ore.
5. Let F be a field, and let R be the free algebra on two letters over F. Prove that R is an integral domain which is neither right Ore nor left Ore.
6. Let R be as in 3.11. Show that R has a classical right quotient ring T and that T is not isomorphic to the maximal right quotient ring of R.
7. Let Q be the maximal right quotient ring of R, and let x be a regular element of R. Prove that x is invertible in Q if and only if $xR \leqq_r R_R$.
8. We say that R satisfies the *right Ore condition* if given $a,x \in R$ with x regular, there exist $b,y \in R$ with y regular such that $ay = xb$. Prove that R has a classical right quotient ring if and only if it satisfies the right Ore condition.
9. If R has a classical right quotient ring and a classical left quotient ring, prove that they coincide.
10. If Q_1 and Q_2 are classical right quotient rings of R, show that the inclusion map $R \to Q_2$ extends to a ring isomorphism of Q_1 onto Q_2.

11. If Q is a classical right quotient ring of R, show that Q is also a classical right quotient ring of itself.
12. If R is commutative and semihereditary, prove that R has a classical right quotient ring which is regular.
13. Prove that any right or left artinian ring is its own classical right quotient ring.
14. Let R be a right Ore domain. Prove that R is left Ore if and only if every finitely generated nonsingular right R-module can be embedded in a free right R-module.
15. Let R be a commutative integral domain with quotient field K, and let T be the ring of all $n \times n$ matrices over K, for some $n > 0$. If P is an R-subalgebra of T such that $\dim(P_R) = n^2$, prove that P is a prime right and left Goldie ring.
16. An element x in a right R-module A is called a *torsion element* if $xr = 0$ for some regular element $r \in R$. Let tA denote the set of all torsion elements of A. Prove that tA is a submodule of A for all A_R if and only if R has a classical right quotient ring.
17. Prove that $tA = Z(A)$ for all A_R if and only if R is a semiprime right Goldie ring.
18. Let R be semiprime right Goldie. A right R-module A is *divisible* if $Ax = A$ for all regular elements $x \in R$. Prove that a nonsingular right R-module is divisible if and only if it is injective.
19. If R is semiprime right Goldie, prove that there is a ring decomposition $R = S \times T$ such that S is semisimple and T has no nonzero injective right ideals. (*Hint*: Exercise 3.B.20.)
20. If R is semiprime right Goldie and $(S^\circ R)_R$ is projective, prove that R is semisimple.

4

Construction Techniques

The purpose of this chapter is to develop several techniques which are useful in constructing examples of nonsingular rings. The text of the chapter is devoted to the basic properties of the rings constructed by these methods and is not concerned with the construction of specific examples. However, many examples are constructed in the exercises, and elsewhere in the book, using the techniques discussed here.

A. Formal Triangular Matrix Rings

Definition Given a pair of rings A,C and a bimodule ${}_C B_A$, we use $\begin{pmatrix} A & 0 \\ B & C \end{pmatrix}$ to denote the set of all symbols $\begin{pmatrix} a & 0 \\ b & c \end{pmatrix}$, where $a \in A$, $b \in B$, $c \in C$. We define operations on $\begin{pmatrix} A & 0 \\ B & C \end{pmatrix}$ according to the usual rules for addition and multiplication of matrices:

$$\begin{pmatrix} a & 0 \\ b & c \end{pmatrix} + \begin{pmatrix} a' & 0 \\ b' & c' \end{pmatrix} = \begin{pmatrix} a + a' & 0 \\ b + b' & c + c' \end{pmatrix}$$

$$\begin{pmatrix} a & 0 \\ b & c \end{pmatrix}\begin{pmatrix} a' & 0 \\ b' & c' \end{pmatrix} = \begin{pmatrix} aa' & 0 \\ ba' + cb' & cc' \end{pmatrix}$$

It is easy to check that $\begin{pmatrix} A & 0 \\ B & C \end{pmatrix}$ is a ring with respect to these operations, and we refer to it as the *formal triangular matrix ring* constructed from A,B,C. For simplicity, we write $R = \begin{pmatrix} A & 0 \\ B & C \end{pmatrix}$ to stand for the hypotheses that A and C are rings, that ${}_C B_A$ is a bimodule, and that R is the formal triangular matrix ring constructed from A,B,C.

For example, the ring of all lower triangular 2×2 matrices over a ring A is the formal triangular matrix ring $\begin{pmatrix} A & 0 \\ A & A \end{pmatrix}$. More generally, if A and C are subrings of a ring B, then $\begin{pmatrix} A & 0 \\ B & C \end{pmatrix}$ is just the subring of the 2×2 matrix ring over B consisting of all matrices of the form $\begin{pmatrix} a & 0 \\ b & c \end{pmatrix}$, where $a \in A$, $b \in B$, $c \in C$. Any ring R can be written trivially as a formal triangular matrix ring, since $R \cong \begin{pmatrix} R & 0 \\ 0 & 0 \end{pmatrix} \cong \begin{pmatrix} 0 & 0 \\ 0 & R \end{pmatrix}$. If R has a nontrivial idempotent e such that $eR(1 - e) = 0$, then R is isomorphic to a nontrivial formal triangular matrix ring $\begin{pmatrix} eRe & 0 \\ (1 - e)Re & (1 - e)R(1 - e) \end{pmatrix}$ (Exercise 1). In case e is central, this is just a direct product decomposition. That is, for any rings A,C we have $A \times C \cong \begin{pmatrix} A & 0 \\ 0 & C \end{pmatrix}$.

Given subsets $X \subseteq A$, $Y \subseteq B$, $Z \subseteq C$, we write

$$\begin{pmatrix} X & 0 \\ Y & Z \end{pmatrix} = \left\{ \begin{pmatrix} x & 0 \\ y & z \end{pmatrix} \middle| \, x \in X, \, y \in Y, \, z \in Z \right\} \subseteq \begin{pmatrix} A & 0 \\ B & C \end{pmatrix}$$

This notation is most useful in case X and Z are one-sided ideals of A and C, and Y is a one-sided submodule of B. For example, if I is a right ideal of A and K is a right ideal of C, then $\begin{pmatrix} I & 0 \\ B & K \end{pmatrix}$ turns out to be a right ideal of $\begin{pmatrix} A & 0 \\ B & C \end{pmatrix}$.

More generally, we need a notation for certain modules over $\begin{pmatrix} A & 0 \\ B & C \end{pmatrix}$. Given a right A-module M, a right C-module N, and a right A-module homomorphism $f: N \otimes_C B \to M$, we construct a right $\begin{pmatrix} A & 0 \\ B & C \end{pmatrix}$-module P as follows. Additively, P is the abelian group $M \oplus N$, but we write each ordered pair (m,n) in the form $\begin{pmatrix} m & \begin{matrix} 0 \\ n \end{matrix} \end{pmatrix}$, which we think of as a 2×2 matrix with left-hand column m and right-hand column $\begin{pmatrix} 0 \\ n \end{pmatrix}$. Multiplication of elements of P by elements of $\begin{pmatrix} A & 0 \\ B & C \end{pmatrix}$ is defined by the rule

$$\begin{pmatrix} m & 0 \\ & n \end{pmatrix}\begin{pmatrix} a & 0 \\ b & c \end{pmatrix} = \begin{pmatrix} ma + f(n \otimes b) & 0 \\ & nc \end{pmatrix}$$

and we check that these definitions make P into a right module over $\begin{pmatrix} A & 0 \\ B & C \end{pmatrix}$. In cases where there is no danger of confusion as to the map f, we write $\begin{pmatrix} M & 0 \\ & N \end{pmatrix}$ for this module P. Actually, all right $\begin{pmatrix} A & 0 \\ B & C \end{pmatrix}$-modules are essentially of this form, as shown in Exercise 2.

In particular, this notation is useful for right ideals of $\begin{pmatrix} A & 0 \\ B & C \end{pmatrix}$. If M is a right A-submodule of $A \oplus B$ and K is a right ideal of C such that $KB \leqq M$, then we get a right $\begin{pmatrix} A & 0 \\ B & C \end{pmatrix}$-module $\begin{pmatrix} M & 0 \\ & K \end{pmatrix}$ by letting $f\colon K \otimes_C B \to M$ be the multiplication map $k \otimes b \mapsto kb$. By identifying $A \oplus B$ with the left-hand column of $\begin{pmatrix} A & 0 \\ B & C \end{pmatrix}$, we may view $\begin{pmatrix} M & 0 \\ & K \end{pmatrix}$ as a right ideal of $\begin{pmatrix} A & 0 \\ B & C \end{pmatrix}$. More precisely,

$$\begin{pmatrix} M & 0 \\ & K \end{pmatrix} = \left\{ \begin{pmatrix} a & 0 \\ b & c \end{pmatrix} \middle| (a,b) \in M,\ c \in K \right\}$$

Similarly, we get left ideals $\begin{pmatrix} I & 0 \\ & N \end{pmatrix}$, where $N \leqq {}_C(B \oplus C)$, I is a left ideal of A, and $BI \leqq N$.

Proposition 4.1 Let $R = \begin{pmatrix} A & 0 \\ B & C \end{pmatrix}$.

(a) If $M \leqq (A \oplus B)_A$, $K \leqq C_C$, and $KB \leqq M$, then $\begin{pmatrix} M & 0 \\ & K \end{pmatrix}$ is a right ideal of R. Conversely, every right ideal of R has this form.

(b) If $I \leqq {}_AA$, $N \leqq {}_C(B \oplus C)$, and $BI \leqq N$, then $\begin{pmatrix} I & 0 \\ & N \end{pmatrix}$ is a left ideal of R. Conversely, every left ideal of R has this form.

(c) If I is a two-sided ideal of A, K a two-sided ideal of C, and J a two-sided submodule of B which contains $BI + KB$, then $\begin{pmatrix} I & 0 \\ J & K \end{pmatrix}$ is a two-sided ideal of R. Conversely, every two-sided ideal of R has this form.

Proof: (a) We have already noted that any such $\begin{pmatrix} M & 0 \\ & K \end{pmatrix}$ is a right ideal of R. Conversely, consider any right ideal P of R. Since $P \cap \begin{pmatrix} A & 0 \\ B & 0 \end{pmatrix}$ is closed under right multiplication by elements of $\begin{pmatrix} A & 0 \\ 0 & 0 \end{pmatrix}$, we must have $P \cap \begin{pmatrix} A & 0 \\ B & 0 \end{pmatrix} = \begin{pmatrix} M & 0 \\ & 0 \end{pmatrix}$ for some $M \leqq (A \oplus B)_A$. Similarly, $P \cap \begin{pmatrix} 0 & 0 \\ 0 & C \end{pmatrix}$ is closed under right multiplication by elements of $\begin{pmatrix} 0 & 0 \\ 0 & C \end{pmatrix}$; hence $P \cap \begin{pmatrix} 0 & 0 \\ 0 & C \end{pmatrix} = \begin{pmatrix} 0 & 0 \\ 0 & K \end{pmatrix}$ for some right ideal K of C. Inasmuch as

$$P = P\begin{pmatrix} 1 & 0 \\ 0 & 0 \end{pmatrix} + P\begin{pmatrix} 0 & 0 \\ 0 & 1 \end{pmatrix} \subseteq P \cap \begin{pmatrix} A & 0 \\ B & 0 \end{pmatrix} + P \cap \begin{pmatrix} 0 & 0 \\ 0 & C \end{pmatrix}$$

we infer that $P = \begin{pmatrix} M & 0 \\ & K \end{pmatrix}$. Finally, since P is closed under right multiplication by elements of $\begin{pmatrix} 0 & 0 \\ B & 0 \end{pmatrix}$, we must have $KB \leqq M$.

(b) follows from (a) by symmetry.

(c) It is clear from (a) and (b) that any such $\begin{pmatrix} I & 0 \\ J & K \end{pmatrix}$ is a two-sided ideal of R. Conversely, let P be any two-sided ideal of R. Since $P \cap \begin{pmatrix} A & 0 \\ 0 & 0 \end{pmatrix}$ is closed under right and left multiplication by elements of $\begin{pmatrix} A & 0 \\ 0 & 0 \end{pmatrix}$, we must have $P \cap \begin{pmatrix} A & 0 \\ 0 & 0 \end{pmatrix} = \begin{pmatrix} I & 0 \\ 0 & 0 \end{pmatrix}$ for some two-sided ideal I of A. Similarly, $P \cap \begin{pmatrix} 0 & 0 \\ 0 & C \end{pmatrix} = \begin{pmatrix} 0 & 0 \\ 0 & K \end{pmatrix}$ for some two-sided ideal K of C, and $P \cap \begin{pmatrix} 0 & 0 \\ B & 0 \end{pmatrix} = \begin{pmatrix} 0 & 0 \\ J & 0 \end{pmatrix}$ for some two-sided submodule J of B. Inasmuch as $\begin{pmatrix} 1 & 0 \\ 0 & 0 \end{pmatrix} R \begin{pmatrix} 0 & 0 \\ 0 & 1 \end{pmatrix} = 0$, we have

$$\begin{aligned} P &= \begin{pmatrix} 1 & 0 \\ 0 & 0 \end{pmatrix} P \begin{pmatrix} 1 & 0 \\ 0 & 0 \end{pmatrix} + \begin{pmatrix} 0 & 0 \\ 0 & 1 \end{pmatrix} P \begin{pmatrix} 1 & 0 \\ 0 & 0 \end{pmatrix} + \begin{pmatrix} 0 & 0 \\ 0 & 1 \end{pmatrix} P \begin{pmatrix} 0 & 0 \\ 0 & 1 \end{pmatrix} \\ &\subseteq P \cap \begin{pmatrix} A & 0 \\ 0 & 0 \end{pmatrix} + P \cap \begin{pmatrix} 0 & 0 \\ B & 0 \end{pmatrix} + P \cap \begin{pmatrix} 0 & 0 \\ 0 & C \end{pmatrix} \end{aligned}$$

from which we obtain $P = \begin{pmatrix} I & 0 \\ J & K \end{pmatrix}$. Finally, since P is closed under right and left multiplication by elements of $\begin{pmatrix} 0 & 0 \\ B & 0 \end{pmatrix}$, we conclude that $BI + KB \leqq J$.□

We now turn to the question of when $R = \begin{pmatrix} A & 0 \\ B & C \end{pmatrix}$ is a nonsingular ring.

This requires information about $\mathscr{S}(R)$ and $Z_r(R)$, which is a bit messy in general. (See Exercise 3.) To simplify things, we restrict ourselves to the case when B is a faithful C-module.

Proposition 4.2 Let $R = \begin{pmatrix} A & 0 \\ B & C \end{pmatrix}$, and assume that ${}_CB$ is faithful.

(a) A right ideal P of R belongs to $\mathscr{S}(R)$ if and only if it contains a right ideal of the form $\begin{pmatrix} I & 0 \\ J & 0 \end{pmatrix}$, where $I \in \mathscr{S}(A)$ and $J \leqq_e B_A$.

(b) $\operatorname{soc}(R_R) = \begin{pmatrix} \operatorname{soc}(A_A) & 0 \\ \operatorname{soc}(B_A) & 0 \end{pmatrix}$.

(c) $Z_r(R) = \begin{pmatrix} Z_r(A) & 0 \\ Z(B_A) & K \end{pmatrix}$, where $K = \{c \in C \mid cJ = 0$ for some $J \leqq_e B_A\}$.

Proof: (a) First assume that $P \in \mathscr{S}(R)$, and write $P = \begin{pmatrix} M & \begin{matrix} 0 \\ K \end{matrix} \end{pmatrix}$ as in 4.1(a). Inasmuch as $P \cap \begin{pmatrix} T & \begin{matrix} 0 \\ 0 \end{matrix} \end{pmatrix} \neq 0$ for all nonzero $T \leqq (A \oplus B)_A$, we obtain $M \cap T \neq 0$ for all such T, that is, $M \leqq_e (A \oplus B)_A$. As a result, $I = M \cap A$ belongs to $\mathscr{S}(A)$, and $J = M \cap B$ is an essential submodule of B_A. Thus $\begin{pmatrix} I & 0 \\ J & 0 \end{pmatrix}$ is a right ideal of R with the required form, and clearly $\begin{pmatrix} I & 0 \\ J & 0 \end{pmatrix} \leqq P$.

For the converse, it suffices to assume that $P = \begin{pmatrix} I & 0 \\ J & 0 \end{pmatrix}$ for some $I \in \mathscr{S}(A)$ and some $J \leqq_e B_A$. Consider any nonzero element $x = \begin{pmatrix} a & 0 \\ b & c \end{pmatrix}$ in R. If $c = 0$, then (a,b) is a nonzero element of $A \oplus B$. Inasmuch as $I \oplus J \leqq_e (A \oplus B)_A$, it follows that $(a,b)A \cap (I \oplus J) \neq 0$, whence $\begin{pmatrix} a & 0 \\ b & 0 \end{pmatrix}\begin{pmatrix} A & 0 \\ 0 & 0 \end{pmatrix} \cap \begin{pmatrix} I & 0 \\ J & 0 \end{pmatrix} \neq 0$. Thus $xR \cap P \neq 0$ in this case. If $c \neq 0$, then since ${}_CB$ is faithful we must have $cB \neq 0$. Now $x\begin{pmatrix} 0 & 0 \\ B & 0 \end{pmatrix} = \begin{pmatrix} 0 & 0 \\ cB & 0 \end{pmatrix}$ is nonzero; hence it follows from the case above that $x\begin{pmatrix} 0 & 0 \\ B & 0 \end{pmatrix} \cap P \neq 0$, and consequently $xR \cap P \neq 0$. Therefore $P \leqq_e R_R$.

(b) According to 1.16, $\operatorname{soc}(R_R) = \cap\mathscr{S}(R)$. In light of (a), we thus obtain $\operatorname{soc}(R_R) = \begin{pmatrix} I & 0 \\ J & 0 \end{pmatrix}$, where $I = \cap\mathscr{S}(A)$ and J is the intersection of all essential submodules of B_A. Using 1.16 again, we conclude that $I = \operatorname{soc}(A_A)$ and $J = \operatorname{soc}(B_A)$.

(c) If $\begin{pmatrix} a & 0 \\ b & c \end{pmatrix} \in Z_r(R)$, then in view of (a) we see that $\begin{pmatrix} a & 0 \\ b & c \end{pmatrix}\begin{pmatrix} I & 0 \\ J & 0 \end{pmatrix} = 0$ for some $I \in \mathscr{S}(A)$ and some $J \leqq_e B_A$. Then $aI = 0$, $bI = 0$, and $cJ = 0$, whence $a \in Z_r(R)$, $b \in Z(B_A)$, and $c \in K$.

Conversely, suppose that $a \in Z_r(A)$, $b \in Z(B_A)$, and $c \in K$. Then $aI_1 = 0$ and $bI_2 = 0$ for some $I_1, I_2 \in \mathscr{S}(A)$; hence we obtain $aI = 0$ and $bI = 0$ for $I = I_1 \cap I_2 \in \mathscr{S}(A)$. Also, $cJ = 0$ for some $J \leqq_e B_A$. Now $\begin{pmatrix} a & 0 \\ b & c \end{pmatrix}\begin{pmatrix} I & 0 \\ J & 0 \end{pmatrix} = 0$, and $\begin{pmatrix} I & 0 \\ J & 0 \end{pmatrix} \in \mathscr{S}(R)$ by (a), whence $\begin{pmatrix} a & 0 \\ b & c \end{pmatrix} \in Z_r(R)$. □

If $R = \begin{pmatrix} A & 0 \\ B & C \end{pmatrix}$ with B faithful on both sides, then we can apply 4.2 as well as the left-right symmetric version, which allows us to identify $\operatorname{soc}(R_R)$, $\operatorname{soc}({}_RR)$, $Z_r(R)$, and $Z_l(R)$. This case can be used to construct examples where R has essential socle on one side but zero socle on the other (as in Exercise 1.C.1), or examples where R is nonsingular on one side but not on the other (as in Exercise 1.D.1).

In general, formal triangular matrix rings are good sources for examples of rings which satisfy some property on one side but not on the other. Exercises 8, 11, 16, and 18 give four such examples.

Corollary 4.3 Let $R = \begin{pmatrix} A & 0 \\ B & C \end{pmatrix}$, and assume that ${}_CB$ is faithful. Then $Z_r(R) = 0$ if and only if $Z_r(A) = 0$ and $Z(B_A) = 0$.

Proof: If $Z_r(R) = 0$, then it is immediate from 4.2 that $Z_r(A) = 0$ and $Z(B_A) = 0$. Conversely, if $Z_r(A) = 0$ and $Z(B_A) = 0$, then 4.2 shows that $Z_r(R) = \begin{pmatrix} 0 & 0 \\ 0 & K \end{pmatrix}$, for a suitable right ideal K of C. Since $Z_r(R)$ is a right ideal of R, 4.1 shows that $KB = 0$. Now the faithfulness of ${}_CB$ implies that $K = 0$, whence $Z_r(R) = 0$. □

Corollary 4.3 can be used to change a situation involving a nonsingular module over a nonsingular ring into a situation involving a right ideal in a nonsingular ring. Specifically, suppose that B is a nonsingular right module over a right nonsingular ring A. According to 4.3, $R = \begin{pmatrix} A & 0 \\ B & \operatorname{End}_A(B) \end{pmatrix}$ is a right nonsingular ring. Also, $\begin{pmatrix} 0 & 0 \\ B & 0 \end{pmatrix}$ is a right ideal of R, and properties of this right ideal correspond directly to properties of the module B_A. For example, $\begin{pmatrix} 0 & 0 \\ B & 0 \end{pmatrix}$ is a finitely generated right ideal of R if and only if B_A is finitely generated. (See Exercise 12.) Alternatively, in order to study the relationship between B and $S^\circ B$, we can take $R = \begin{pmatrix} A & 0 \\ S^\circ B & \mathrm{End}_A(S^\circ B) \end{pmatrix}$ and study the relationship between the right ideals $\begin{pmatrix} 0 & 0 \\ B & 0 \end{pmatrix}$ and $\begin{pmatrix} 0 & 0 \\ S^\circ B & 0 \end{pmatrix}$.

In case we have a formal triangular matrix ring $R = \begin{pmatrix} A & 0 \\ B & C \end{pmatrix}$ which is nonsingular, we would like to find $S^\circ R$. As in 4.2, we can simplify things by assuming that ${}_C B$ is faithful, in which case we make two normalizations. First, since C acts faithfully on B, we might as well assume that C is a subring of $\mathrm{End}_A(B)$. Inasmuch as B_A is nonsingular (by 4.3), each $f \in C$ extends uniquely to an endomorphism of $S^\circ B$, namely, $S^\circ f$. Thus, the functor S° provides an isomorphism $f \mapsto S^\circ f$ of C onto a subring of $\mathrm{End}_A(S^\circ B)$. As the second normalization, we may thus assume that C is a subring of $\mathrm{End}_A(S^\circ B)$ for which $CB \leqq B$.

Proposition 4.4 Let $R = \begin{pmatrix} A & 0 \\ B & C \end{pmatrix}$, where $Z_r(A) = 0$, $Z(B_A) = 0$, and C is a subring of $\mathrm{End}_A(S^\circ B)$ such that $CB \leqq B$. Then $Z_r(R) = 0$ and

$$S^\circ R = \begin{pmatrix} S^\circ A & \mathrm{Hom}_A(S^\circ B, S^\circ A) \\ S^\circ B & \mathrm{End}_A(S^\circ B) \end{pmatrix}$$

[The product bf of an element $b \in S^\circ B$ with a map $f \in \mathrm{Hom}_A(S^\circ B, S^\circ A)$ is defined to be the map $x \mapsto b(fx)$ in $\mathrm{End}_A(S^\circ B)$.]

Proof: Set $Q = \begin{pmatrix} S^\circ A & \mathrm{Hom}_A(S^\circ B, S^\circ A) \\ S^\circ B & \mathrm{End}_A(S^\circ B) \end{pmatrix}$, and note that Q is a ring containing R as a subring. We proceed via 2.11, which requires us to show that $R_R \leqq_e Q_R$ and that Q is a regular right self-injective ring.

Inasmuch as all right A-endomorphisms of $S^\circ A$ are also right $S^\circ A$-endomorphisms, we may identify $\mathrm{End}_A(S^\circ A)$ with $S^\circ A$. Likewise, we may identify $\mathrm{Hom}_A(S^\circ A, S^\circ B)$ with $S^\circ B$, at which point we infer that

$$Q \cong \begin{pmatrix} \mathrm{End}_A(S^\circ A) & \mathrm{Hom}_A(S^\circ B, S^\circ A) \\ \mathrm{Hom}_A(S^\circ A, S^\circ B) & \mathrm{End}_A(S^\circ B) \end{pmatrix} \cong \mathrm{End}_A(S^\circ A \oplus S^\circ B).$$

Since $S^\circ A \oplus S^\circ B$ is a nonsingular injective right A-module, 2.22 now shows that Q is regular and right self-injective.

Now consider any nonzero element $x = \begin{pmatrix} a & f \\ b & g \end{pmatrix} \in Q$. If $f = 0$ and $g = 0$, then (a,b) is a nonzero element of $S^\circ A \oplus S^\circ B$. Since $A \oplus B \leqq_e S^\circ A \oplus S^\circ B$, we obtain $(a,b)A \cap (A \oplus B) \neq 0$; hence $\begin{pmatrix} a & 0 \\ b & 0 \end{pmatrix}\begin{pmatrix} A & 0 \\ 0 & 0 \end{pmatrix} \cap \begin{pmatrix} A & 0 \\ B & 0 \end{pmatrix} \neq 0$.

Consequently, $xR \cap R \neq 0$ in this case. If $f \neq 0$, then it follows from 2.1 that $fB \neq 0$, whence $x\begin{pmatrix} 0 & 0 \\ B & 0 \end{pmatrix} \leqq \begin{pmatrix} fB & 0 \\ gB & 0 \end{pmatrix}$ is nonzero. It now follows from the case above that $x\begin{pmatrix} 0 & 0 \\ B & 0 \end{pmatrix} \cap R \neq 0$; hence $xR \cap R \neq 0$. Finally, if $g \neq 0$ we similarly obtain $gB \neq 0$ and then $xR \cap R \neq 0$. Therefore $R_R \leqq_e Q_R$.

According to 2.11, we now have $Z_r(R) = 0$ and $S^\circ R = Q$.□

For example, 4.4 applies to the case when $Z_r(A) = 0$ and $R = \begin{pmatrix} A & 0 \\ S^\circ A & S^\circ A \end{pmatrix}$; then $Z_r(R) = 0$ and $S^\circ R$ is just the ring of all 2×2 matrices over $S^\circ A$. Exercise 9 shows one of the possible connections between $S^\circ R$ and $S^\circ A$ in this situation.

We conclude this section by deriving necessary and sufficient conditions under which a formal triangular matrix ring is hereditary. We proceed in a relatively straightforward manner, but with some handwaving. More precision can be obtained (with a corresponding increase in obscurity) by using the functors defined in Exercises 19 and 20.

Proposition 4.5 Let $P = \begin{pmatrix} M & \begin{matrix} 0 \\ K \end{matrix} \end{pmatrix}$ be any right ideal in the ring $R = \begin{pmatrix} A & 0 \\ B & C \end{pmatrix}$. Then P is projective if and only if $(M/KB)_A$ and K_C are projective and the map $K \otimes_C B \to B$ is a monomorphism.

Proof: First assume that P_R is projective. Since $H = \begin{pmatrix} 0 & 0 \\ B & C \end{pmatrix}$ is a two-sided ideal of R, it follows that P/PH is a projective right (R/H)-module. Observing that $P/PH = \begin{pmatrix} M & \begin{matrix} 0 \\ K \end{matrix} \end{pmatrix} \Big/ \begin{pmatrix} KB & \begin{matrix} 0 \\ K \end{matrix} \end{pmatrix}$ and that $R/H \cong A$, we infer that $(M/KB)_A$ must be projective. Similarly, $(P/PL)_{R/L}$ is projective, where $L = \begin{pmatrix} A & 0 \\ B & 0 \end{pmatrix}$. Inasmuch as $P/PL = \begin{pmatrix} M & \begin{matrix} 0 \\ K \end{matrix} \end{pmatrix} \Big/ \begin{pmatrix} M & \begin{matrix} 0 \\ 0 \end{matrix} \end{pmatrix}$ and $R/L \cong C$, we see that K_C is projective.

Since P_R is flat and $E = \begin{pmatrix} 0 & 0 \\ B & 0 \end{pmatrix}$ is a left ideal of R, the map $P \otimes_R E \to P \to R$ must be a monomorphism. Now $E = \begin{pmatrix} 0 & 0 \\ 0 & 1 \end{pmatrix}E$ and $P\begin{pmatrix} 0 & 0 \\ 0 & 1 \end{pmatrix} = \begin{pmatrix} 0 & 0 \\ 0 & K \end{pmatrix}$, from which we infer that $P \otimes_R E$ is naturally isomorphic to $K \otimes_C B$. It follows that the map $K \otimes_C B \to B$ is a monomorphism.

Conversely, assume that $(M/KB)_A$ and K_C are projective, and that the map $K \otimes_C B \to B$ is a monomorphism. Setting $H = \begin{pmatrix} 0 & 0 \\ B & C \end{pmatrix}$, we note as above that $P/PH = \begin{pmatrix} M & 0 \\ & K \end{pmatrix} \Big/ \begin{pmatrix} KB & 0 \\ & K \end{pmatrix}$ and $R/H \cong A$, from which we infer that $(P/PH)_{R/H}$ is projective. Since H_R is generated by the idempotent $\begin{pmatrix} 0 & 0 \\ 0 & 1 \end{pmatrix}$, we see that $(R/H)_R$ is projective, from which it follows that P/PH is also projective as a right R-module.

Inasmuch as K_C is projective, there is a decomposition $K \oplus K' \cong \oplus C_\alpha$ of right C-modules such that each $C_\alpha \cong C_C$. Consequently,

$$\begin{pmatrix} K \otimes_C B & 0 \\ & K \end{pmatrix} \oplus \begin{pmatrix} K' \otimes_C B & 0 \\ & K' \end{pmatrix} \cong \oplus \begin{pmatrix} C_\alpha \otimes_C B & 0 \\ & C_\alpha \end{pmatrix}$$

which is isomorphic to a direct sum of copies of H. We have already noted that H_R is generated by an idempotent, whence H_R is projective, and so $\begin{pmatrix} K \otimes_C B & 0 \\ & K \end{pmatrix}$ is a projective right R-module. Since the map $K \otimes_C B \to B$ is a monomorphism, we see that $\begin{pmatrix} K \otimes_C B & 0 \\ & K \end{pmatrix}$ is isomorphic to $\begin{pmatrix} KB & 0 \\ & K \end{pmatrix} = PH$; hence $(PH)_R$ must be projective.

Therefore P/PH and PH are both projective right R-modules, whence P is projective.□

Lemma 4.6 Let $L \leqq H$ be modules. If every submodule of H which is comparable to L is projective, then every submodule of H is projective.

Proof: Given any $G \leqq H$, our hypothesis implies that $G + L$ and $G \cap L$ are both projective. Since there exists an exact sequence $0 \to G \cap L \to G \oplus L \to G + L \to 0$, we conclude that $G \oplus L$ must be projective, whence G is projective.□

Theorem 4.7 The ring $R = \begin{pmatrix} A & 0 \\ B & C \end{pmatrix}$ is right hereditary if and only if

(a) A and C are both right hereditary.
(b) ${}_C B$ is flat.
(c) $(B/KB)_A$ is projective for all $K \leqq C_C$.

Proof: First assume that R is right hereditary. Given any right ideal I of A, $\begin{pmatrix} I & 0 \\ 0 & 0 \end{pmatrix}$ is a right ideal of R and so is projective, whence 4.5 says that I_A is projective. Therefore A is right hereditary. Given any right ideal K of C, $\begin{pmatrix} 0 & 0 \\ B & K \end{pmatrix}$ is a right ideal of R and thus is projective; hence 4.5 says that $(B/KB)_A$ and K_C are projective and that the map $K \otimes_C B \to B$ is a monomorphism. Therefore C is right hereditary and ${}_C B$ is flat.

Conversely, assume that (a) to (c) hold. We first apply 4.6 to the right ideals $L = \begin{pmatrix} 0 & 0 \\ B & 0 \end{pmatrix}$ and $H = \begin{pmatrix} 0 & 0 \\ B & C \end{pmatrix}$. Given any $P \leqq L_R$, we must have $P = \begin{pmatrix} 0 & 0 \\ M & 0 \end{pmatrix}$ for some $M \leqq B_A$. Since B_A is projective and A is right hereditary, M_A is projective, whence 4.5 says that P_R is projective. Now consider any $P \leqq H_R$ which contains L, in which case $P = \begin{pmatrix} 0 & 0 \\ B & K \end{pmatrix}$ for some right ideal K of C. Then $(B/KB)_A$ is projective by (c), K_C is projective because C is right hereditary, and $K \otimes_C B \to B$ is a monomorphism because ${}_C B$ is flat; hence 4.5 shows that P_R is projective. Thus all submodules of H_R comparable to L are projective; hence 4.6 says that all submodules of H_R are projective.

We conclude by applying 4.6 to the pair $H_R \leqq R_R$. Given any right ideal P of R which contains H, we have $P = \begin{pmatrix} I & 0 \\ B & C \end{pmatrix}$ for some right ideal I of A. Now $(I \oplus B)/CB \cong I$ is projective because A is right hereditary, C_C is certainly projective, and the map $C \otimes_C B \to B$ is an isomorphism; hence 4.5 says that P_R is projective. Therefore all right ideals of R comparable to H are projective, and consequently 4.6 says that all right ideals of R are projective. □

Corollary 4.8 Let $R = \begin{pmatrix} A & 0 \\ B & C \end{pmatrix}$. If A is right hereditary, B_A is projective, and C is semisimple, then R is right hereditary.

Proof: Certainly A and C are both right hereditary. Since C is semisimple, ${}_C B$ is projective and thus flat. Any right ideal K of C must have the form eC for some idempotent $e \in C$, whence $KB = eB$ is a direct summand of B_A. Then $(B/KB)_A$ is isomorphic to a direct summand of the projective module B_A and so is projective. Now 4.7 shows that R is right hereditary. □

Corollary 4.9 Let T be a semisimple ring. For any positive integer n, the ring R_n of all lower triangular $n \times n$ matrices over T is right and left hereditary.

Proof: By symmetry, it suffices to prove that R_n is right hereditary. Since $R_1 \cong T$, R_1 is certainly right hereditary.

Now let $n > 1$, and assume that R_{n-1} is right hereditary. We identify R_n with $\begin{pmatrix} R_{n-1} & 0 \\ B & T \end{pmatrix}$, where B is isomorphic as a right R_{n-1}-module to the bottom row of R_{n-1}. Inasmuch as R_{n-1} is right hereditary, B must be a projective right R_{n-1}-module; hence 4.8 shows that R_n is right hereditary. □

Exercises

1. If $e = e^2 \in R$ and $eR(1-e) = 0$, show that $R \cong \begin{pmatrix} eRe & 0 \\ (1-e)Re & (1-e)R(1-e) \end{pmatrix}$.

2. Let $R = \begin{pmatrix} A & 0 \\ B & C \end{pmatrix}$, and define a category $\mathcal{M}$ as follows. The objects of $\mathcal{M}$ are all triples (M,N,f), where M is a right A-module, N is a right C-module, and $f : N \otimes_C B \to M$ is a right A-homomorphism. A morphism from (M,N,f) to (M',N',f') in $\mathcal{M}$ is any pair (g,h), where $g : M \to M'$ is an A-homomorphism, $h : N \to N'$ is a C-homomorphism, and $gf = f'(h \otimes 1)$. Show that $\mathcal{M}$ is a category equivalent to Mod-R.

3. Let $R = \begin{pmatrix} A & 0 \\ B & C \end{pmatrix}$, and set $H = \{c \in C \mid cB = 0\}$. Show that a right ideal of R belongs to $\mathcal{S}(R)$ if and only if it contains a right ideal of the form $\begin{pmatrix} I & 0 \\ J & K \end{pmatrix}$, where $I \in \mathcal{S}(A)$, $J \leqq_e B_A$, and $K \leqq_e H_C$. Show that
$$\mathrm{soc}(R_R) = \begin{pmatrix} \mathrm{soc}(A_A) & 0 \\ \mathrm{soc}(B_A) & \mathrm{soc}(H_C) \end{pmatrix}.$$
Show that $Z_r(R) = \begin{pmatrix} Z_r(A) & 0 \\ Z(B_A) & L \cap L' \end{pmatrix}$, where $L = \{c \in C \mid cJ = 0$ for some $J \leqq_e B_A\}$ and $L' = \{c \in C \mid cK = 0$ for some $K \leqq_e H_C\}$.

4. Let $R = \begin{pmatrix} A & 0 \\ B & C \end{pmatrix}$ as in 4.4, and set $H = \begin{pmatrix} 0 & 0 \\ B & C \end{pmatrix}$. Show that H_R is injective if and only if B_A is injective and $C = \{f \in \mathrm{End}_A(S^\circ B) \mid fB \leqq B\}$.

5. Let $Z_r(R) = 0$ and assume that R_R is finite-dimensional. Prove that $R \cong \begin{pmatrix} A & 0 \\ B & C \end{pmatrix}$ such that $Z_r(A) = 0$, B_A is nonsingular injective, ${}_CB$ is faithful, C is semisimple, and all injective right ideals of $\begin{pmatrix} A & 0 \\ B & C \end{pmatrix}$ are contained in $\begin{pmatrix} 0 & 0 \\ B & C \end{pmatrix}$. (*Hint*: Exercise 3.B.20.)

6. Show that any algebra over a field can be expressed as a factor ring of a right nonsingular ring.

7. Let $R = \begin{pmatrix} A & 0 \\ B & C \end{pmatrix}$ with ${}_CB$ faithful. Show that R_R is finite-dimensional if and only if A_A and B_A are both finite-dimensional, in which case $\dim(R_R) = \dim(A_A) + \dim(B_A)$.

8. Let $F \subseteq K$ be fields such that $[K\colon F] = \infty$, and set $R = \begin{pmatrix} K & 0 \\ K & F \end{pmatrix}$. Show that R_R is finite-dimensional but that ${}_RR$ is not.

9. Let $Z_r(A) = 0$ and set $R = \begin{pmatrix} A & 0 \\ S^\circ A & S^\circ A \end{pmatrix}$. If ${}_R(S^\circ R)$ is flat, prove that ${}_A(S^\circ A)$ is flat and that the multiplication map $S^\circ A \otimes_A S^\circ A \to S^\circ A$ is an isomorphism. (Compare Exercise 3.A.11.)

10. Let A be a commutative hereditary ring. If B is any nonsingular semisimple A-module, show that $\begin{pmatrix} A & 0 \\ B & A \end{pmatrix}$ is right and left hereditary.

11. Show that $\begin{pmatrix} \mathbf{Q} & 0 \\ \mathbf{Q} & \mathbf{Z} \end{pmatrix}$ is right hereditary but not left hereditary.

12. Show that a right ideal P of $\begin{pmatrix} A & 0 \\ B & C \end{pmatrix}$ is finitely generated if and only if $P = \begin{pmatrix} M + KB & 0 \\ & K \end{pmatrix}$ for some finitely generated $M \leqq (A \oplus B)_A$ and some finitely generated $K \leqq C_C$.
13. Let $R = \begin{pmatrix} A & 0 \\ B & C \end{pmatrix}$. If A is right semihereditary, B_A is projective, and C is regular, prove that R is right semihereditary.
14. If T is a regular ring and $n > 0$, prove that the ring of all $n \times n$ lower triangular matrices over T is right and left semihereditary.
15. Show that $\begin{pmatrix} A & 0 \\ B & C \end{pmatrix}$ is right noetherian if and only if A and C are right noetherian and B_A is finitely generated.
16. Show that $\begin{pmatrix} \mathbf{Q} & 0 \\ \mathbf{Q} & \mathbf{Z} \end{pmatrix}$ is right noetherian but not left noetherian.
17. Show that $\begin{pmatrix} A & 0 \\ B & C \end{pmatrix}$ is right artinian if and only if A and C are right artinian and B_A is finitely generated.
18. Show that the ring R of Exercise 8 is right artinian and right noetherian, but neither left artinian nor left noetherian.
19. Let $R = \begin{pmatrix} A & 0 \\ B & C \end{pmatrix}$. Define functors $J_1 : \text{Mod-}A \to \text{Mod-}R$ and J_2, J_{23}, J_3 : Mod-$C \to$ Mod-R as follows: $J_1(M) = \begin{pmatrix} M & 0 \\ & 0 \end{pmatrix}$, $J_2(N) = \begin{pmatrix} N \otimes_C B & 0 \\ & 0 \end{pmatrix}$, $J_{23}(N) = \begin{pmatrix} N \otimes_C B & 0 \\ & N \end{pmatrix}$, and $J_3(N) = J_{23}(N)/J_2(N)$. Show that all four functors are additive and preserve arbitrary direct sums. Show that J_1 and J_3 are exact, while J_2 and J_{23} are exact if and only if ${}_C B$ is flat. Show that J_1 and J_{23} preserve projectives.
20. Let $R = \begin{pmatrix} A & 0 \\ B & C \end{pmatrix}$. Define functors $P_1, P_{12}, P_2 : \text{Mod-}R \to \text{Mod-}A$ and P_3 : Mod-$R \to$ Mod-C as follows: $P_2(M) = M\begin{pmatrix} 0 & 0 \\ B & 0 \end{pmatrix}$, $P_{12}(M) = M\begin{pmatrix} A & 0 \\ B & 0 \end{pmatrix}$, $P_1(M) = P_{12}(M)/P_2(M)$, and $P_3(M) = M\begin{pmatrix} 0 & 0 \\ 0 & C \end{pmatrix}$. Show that all four functors are additive and preserve arbitrary direct sums. Show that P_{12} and P_3 are exact, while P_1 and P_2 are exact if and only if $B = 0$. Show that P_1 and P_3 preserve projectives.
21. With notation as in Exercises 19 and 20, show that $P_{12}J_1$ and P_1J_1 are naturally equivalent to the identity functor on Mod-A and that P_3J_{23} and P_3J_3 are naturally equivalent to the identity functor on Mod-C.
22. With notation as in Exercises 19 and 20, show that (J_1, P_{12}), (J_{23}, P_3), (P_1, J_1), and (P_3, J_3) are adjoint pairs of functors.

B. Essential Products

Definition A *subdirect product* of a collection $\{R_\gamma\}$ of rings is any subring R of ΠR_γ such that each of the projections $R \to \Pi R_\gamma \to R_\lambda$ is surjective. Obviously ΠR_γ qualifies as a subdirect product of the R_γ. Also, any subring of ΠR_γ which contains $\oplus R_\gamma$ is a subdirect product of the R_γ. In general, a ring R is isomorphic to a subdirect product of rings R_γ if and only if R contains a two-sided ideal H_γ for each γ such that $R/H_\gamma \cong R_\gamma$ for each γ and $\cap H_\gamma = 0$ (Exercise 1).

Definition A *right essential product* of a collection $\{R_\gamma\}$ of rings is any subdirect product of the R_γ which contains an essential right ideal of the ring ΠR_γ. Clearly ΠR_γ is a right essential product of the R_γ. For another example, set $R_n = \mathbf{Z}$ for $n = 1,2,\ldots$ and define

$$R = \{x \in \Pi R_n \mid x_i \equiv x_j \pmod 2 \text{ for all } i,j\}$$

It is clear that R is a subdirect product of the R_n, and R contains $\Pi(2R_n)$, which is an essential right ideal of ΠR_n. Thus R is a right essential product of the R_n. We note that the concept of an essential product is definitely stronger than the concept of a subdirect product (Exercise 2). Also, the concept of an essential product is not left-right symmetric, as shown by Exercise 3.

Essential products occur many places where direct products do not. For example, every finite-dimensional nonsingular semiprime ring is isomorphic to an essential product of prime rings (Exercise 9), but such a ring need not be isomorphic to a direct product of prime rings (Exercise 10). Similarly, Theorem 4.17 shows that every nonsingular ring is an essential product of a ring with essential socle and a ring with zero socle, but there exist nonsingular rings which are not isomorphic to a direct product of a ring with essential socle and a ring with zero socle (Exercise 12).

We now consider several alternate descriptions of essential products. In order to simplify the notation, we identify any factor R_λ of a direct product ΠR_γ with the ring

$$\{x \in \Pi R_\gamma \mid x_\gamma = 0 \text{ for all } \gamma \neq \lambda\}$$

(This is *not* a subring of ΠR_γ, because of our convention that subrings of a ring must have the same unit.) Given any subring R of ΠR_γ, it thus makes sense to talk about $R \cap R_\lambda$, which is at least a two-sided ideal of R.

Proposition 4.10 Let R be a subdirect product of rings R_γ.

(a) For each λ, $R \cap R_\lambda$ is a two-sided ideal of R_λ.

(b) R is a right essential product of the R_γ if and only if $R \cap R_\lambda \in \mathscr{S}(R_\lambda)$ for all λ.

Proof: (a) Given $x \in R \cap R_\lambda$ and $s \in R_\lambda$, we have $x_\gamma = s_\gamma = 0$ for all $\gamma \neq \lambda$. Since R is a subdirect product of the R_γ, there must exist an element $r \in R$ for which $r_\lambda = s_\lambda$. Observing that $(xr)_\gamma = 0 = (xs)_\gamma$ for all $\gamma \neq \lambda$, we see that $xr = xs$, whence $xs \in R \cap R_\lambda$. Thus $R \cap R_\lambda$ is a right ideal of R_λ, and by symmetry it is a left ideal as well.

(b) If $R \cap R_\lambda \in \mathscr{S}(R_\lambda)$ for all λ, then $(R \cap R_\lambda)_T \leqq_e (R_\lambda)_T$ for all λ, where $T = \Pi R_\gamma$. Consequently, $[\oplus(R \cap R_\gamma)]_T \leqq_e (\oplus R_\gamma)_T$. Observing that $(\oplus R_\gamma)_T \leqq_e T_T$, we thus find that $\oplus(R \cap R_\lambda) \in \mathscr{S}(T)$. Inasmuch as $\oplus(R \cap R_\lambda) \subseteq R$, it follows that R is a right essential product of the R_γ.

Conversely, assume that R is a right essential product of the R_γ. Now R must contain some J which belongs to $\mathscr{S}(\Pi R_\gamma)$, and we claim that for any λ, $J \cap R_\lambda \in \mathscr{S}(R_\lambda)$. Given any nonzero $x \in R_\lambda$, we at least have an element $y \in \Pi R_\gamma$ such that xy is a nonzero element of J. Since $x_\gamma = 0$ for all $\gamma \neq \lambda$, the only way $xy \neq 0$ is to have $x_\lambda y_\lambda \neq 0$; hence xy_λ must be a nonzero element of $J \cap R_\lambda$. Thus $J \cap R_\lambda \in \mathscr{S}(R_\lambda)$. Inasmuch as $R \cap R_\lambda$ is a right ideal of R_λ which contains $J \cap R_\lambda$, we conclude that $R \cap R_\lambda \in \mathscr{S}(R_\lambda)$.□

Theorem 4.11 Let R be a subring of a direct product $R_1 \times R_2$. Then R is a right essential product of R_1 and R_2 if and only if each R_i has a two-sided ideal E_i which belongs to $\mathscr{S}(R_i)$ and there exists a ring isomorphism $\phi: R_1/E_1 \to R_2/E_2$ such that

$$R = \{(x_1,x_2) \in R_1 \times R_2 \mid \phi(x_1 + E_1) = x_2 + E_2\}$$

Proof: First assume that R can be expressed in such a way. Given any $x_1 \in R_1$, there must exist an element $x_2 \in R_2$ such that $x_2 + E_2 = \phi(x_1 + E_1)$; hence (x_1,x_2) is an element of R whose first component is x_1. Similarly, any element of R_2 can be found as the second component of a suitable element of R, whence R is a subdirect product of R_1 and R_2. Inasmuch as each $E_i \in \mathscr{S}(R_i)$, we must have $E_1 \times E_2 \in \mathscr{S}(R_1 \times R_2)$. Given any $(e_1,e_2) \in E_1 \times E_2$, we have $\phi(e_1 + E_1) = 0 = e_2 + E_2$, whence $(e_1,e_2) \in R$. Thus $E_1 \times E_2 \subseteq R$, and consequently R is a right essential product of R_1 and R_2.

Now assume that R is a right essential product of R_1 and R_2, and note from 4.10 that each $E_i = R \cap R_i$ is a two-sided ideal of R_i which belongs to $\mathscr{S}(R_i)$. Inasmuch as R is a subdirect product of R_1 and R_2, the projection $R_1 \times R_2 \to R_1$ induces an isomorphism of R/E_2 onto R_1, which in turn induces an isomorphism $f_1 : R/(E_1 \oplus E_2) \to R_1/E_1$. Likewise, the projection $R_1 \times R_2 \to R_2$ induces an isomorphism $f_2 : R/(E_1 \oplus E_2) \to R_2/E_2$, from which we obtain an isomorphism $\phi = f_2 f_1^{-1}: R_1/E_1 \to R_2/E_2$.

Given any element $r = (x_1,x_2) \in R$, we have $f_i(r + E_1 \oplus E_2) = x_i + E_i$ for each i, whence $\phi(x_1 + E_1) = x_2 + E_2$. Now consider any element $(x_1,x_2) \in R_1 \times R_2$ such that $\phi(x_1 + E_1) = x_2 + E_2$. Then $x_1 + E_1 = f_1(r + E_1 \oplus E_2)$ for some $r \in R$, from which we obtain $x_2 + E_2 = f_2(r + E_1 \oplus E_2)$ as well. Consequently, $r = (x_1 + e_1, x_2 + e_2)$ for some $e_i \in E_i$, from which we conclude that $(x_1,x_2) = r - (e_1,e_2)$ belongs to R. Therefore R has the required form.□

The procedure described in 4.11 can also be used to construct right essential products with more than two factors (Exercise 4), but not all right essential products can be obtained in this way (Exercise 5).

Proposition 4.12 Let R be a subdirect product of a finite collection $\{R_1, \ldots, R_n\}$ of rings. Then R is a right essential product of the R_i if and only if $R_R \leqq_e (\Pi R_i)_R$.

Proof: For each i, set $E_i = R \cap R_i$, which is a two-sided ideal of R_i by 4.10.

If $R_R \leqq_e (\Pi R_i)_R$, then $(E_j)_R = (R \cap R_j)_R \leqq_e (R_j)_R$ for each j. As a consequence, we infer that each $E_j \in \mathcal{S}(R_j)$, whence 4.10 says that R is a right essential product of the R_i.

Conversely, assume that R is a right essential product of the R_i. Since R is a subdirect product of the R_i, we have $p_jR = R_j$ for each projection $p_j: \Pi R_i \to R_j$, from which we infer that the right R-submodules of R_j are exactly the right ideals of R_j. Since $E_j \in \mathcal{S}(R_j)$ by 4.10, we thus see that $(E_j)_R \leqq_e (R_j)_R$, whence

$$(E_1 \oplus \cdots \oplus E_n)_R \leqq_e (R_1 \oplus \cdots \oplus R_n)_R$$

We also have $\oplus E_i \subseteq R$, from which we conclude that $R_R \leqq_e (\Pi R_i)_R$.□

The results of 4.12 may fail in cases with infinitely many factors (Exercise 6), but this problem can be avoided by restricting attention to nonsingular rings, as the next theorem shows.

Theorem 4.13 If R is a subdirect product of a collection $\{R_\gamma\}$ of right nonsingular rings, then the following conditions are equivalent:

(a) R is a right essential product of the R_γ.

(b) $R_R \leqq_e (\Pi R_\gamma)_R$.

(c) ΠR_γ is a right quotient ring of R.

Proof: For each γ, set $E_\gamma = R \cap R_\gamma$, which is a two-sided ideal of R_γ by 4.10.

(c) ⇒ (b) is automatic.

(b) ⇒ (a): In view of (b), we have $(E_\lambda)_R = (R \cap R_\lambda)_R \leqq_e (R_\lambda)_R$ for each λ. Then $E_\lambda \in \mathcal{S}(R_\lambda)$ for all λ, and 4.10 shows that R is a right essential product of the R_γ.

(a) ⇒ (c): For each γ, $E_\gamma \in \mathcal{S}(R_\gamma)$ by 4.10, and $Z_r(R_\gamma) = 0$ by hypothesis; hence the left annihilator of E_γ in R_γ must be zero. Thus $\oplus E_\gamma$ is a two-sided ideal of ΠR_γ whose left annihilator is zero.

Given any $x,y \in \Pi R_\gamma$ with $x \neq 0$, we must have $xe \neq 0$ for some $e \in \oplus E_\gamma$, and since $\oplus E_\gamma$ is a two-sided ideal of ΠR_γ we also obtain $ye \in \oplus E_\gamma$. Inasmuch as $\oplus E_\gamma \subseteq R$, we thus have an element $e \in R$ such that $xe \neq 0$ and $ye \in R$. According to 2.25, $R_R \leqq_r (\Pi R_\gamma)_R$, whence ΠR_γ is a right quotient ring of R.□

Proposition 4.14 Let R be a right essential product of right nonsingular rings R_γ, and set $T = \Pi R_\gamma$.

(a) $\mathscr{S}(T) = \{J \leqq T_T \mid J \cap R \in \mathscr{S}(R)\}$.

(b) $\mathscr{S}(R) = \{K \leqq R_R \mid KT \in \mathscr{S}(T)\}$.

(c) $Z_T(A) = Z_R(A)$ for all A_T.

(d) $Z_r(R) = Z(T_R) = 0$.

Proof: Since T is a right quotient ring of R (by 4.13), (a) to (c) follow directly from 2.32.

(d) Given any $I \in \mathscr{S}(T)$, it is clear that $I \cap R_\gamma \in \mathscr{S}(R_\gamma)$ for all γ. Inasmuch as each $Z_r(R_\gamma) = 0$, the left annihilator of $I \cap R_\gamma$ in R_γ must be zero, whence the left annihilator of $\oplus(I \cap R_\gamma)$ in T is zero. Since $\oplus(I \cap R_\gamma) \subseteq I$, the left annihilator of I must be zero also. Therefore $Z_r(T) = 0$, whence (c) shows that $Z(T_R) = 0$, and consequently $Z(R_R) = 0$.□

Exercise 7 gives two other descriptions of $\mathscr{S}(R)$ in the situation covered by 4.14. We note that 4.14(a) to (c) may fail if the rings R_γ are not nonsingular (Exercise 8).

Proposition 4.15 If R is a right essential product of right nonsingular rings R_γ, then $Z_r(R) = 0$ and $S^\circ R = \Pi(S^\circ R_\gamma)$.

Proof: Since each of the rings $S^\circ R_\gamma$ is regular and right self-injective, the same must hold for the ring $Q = \Pi(S^\circ R_\gamma)$. If $T = \Pi R_\gamma$, then $Z_r(T) = 0$ by 4.14, and we infer that $T_T \leqq_e Q_T$. It follows from 2.27 that Q is a right quotient ring of T. Inasmuch as T is a right quotient ring of R by 4.13, we see from 2.28 that Q is a right quotient ring of R. Consequently, $R_R \leqq_e Q_R$; hence 2.11 says that $Z_r(R) = 0$ and $S^\circ R = Q$.□

With the help of 4.14, we now derive some criteria which tell when a ring is isomorphic to an essential product of nonsingular rings.

Theorem 4.16 Let $\{R_\gamma \mid \gamma \in X\}$ be a collection of right nonsingular rings. A ring R is isomorphic to a right essential product of the R_γ if and only if R contains two-sided ideals $\{H_\gamma \mid \gamma \in X\}$ such that

(a) $R/H_\gamma \cong R_\gamma$ for all γ.

(b) $\cap H_\gamma = 0$.
(c) Each $H_\gamma \in L^*(R_R)$.
(d) $\oplus_\gamma(\bigcap_{\lambda \neq \gamma} H_\lambda) \in \mathscr{S}(R)$.

Proof: First assume we have a ring isomorphism ϕ of R onto some right essential product P of the R_γ. For each λ, set $K_\lambda = P \cap (\prod_{\gamma \neq \lambda} R_\gamma)$, which is a two-sided ideal of P. Inasmuch as P is a subdirect product of the R_γ, each projection $\Pi R_\gamma \to R_\lambda$ induces an isomorphism of P/K_λ onto R_λ. Since $\bigcap_\lambda (\prod_{\gamma \neq \lambda} R_\gamma) = 0$, we obtain $\cap K_\lambda = 0$. According to 4.14, ΠR_γ is a nonsingular right P-module; hence the same is true of any factor R_λ. Observing that the ring isomorphism $P/K_\lambda \to R_\lambda$ is also a right P-module isomorphism, we see that $(P/K_\lambda)_P$ is nonsingular, that is, $K_\lambda \in L^*(P_P)$. Finally, set $K = \oplus_\gamma(\bigcap_{\lambda \neq \gamma} K_\lambda)$, and note that $K = \oplus(\mathrm{P} \cap \mathrm{R}_\gamma)$. Since each $P \cap R_\gamma \in \mathscr{S}(R_\gamma)$ by 4.10, we infer that $K \in \mathscr{S}(\Pi R_\gamma)$; hence 4.14 shows that $K = K \cap P \in \mathscr{S}(P)$. Therefore $\{H_\gamma\} = \{\phi^{-1} K_\gamma\}$ is the required set of two-sided ideals of R.

Conversely, assume that R has two-sided ideals H_γ satisfying (a) to (d). For each γ, let ϕ_γ be the composition of the natural map $R \to R/H_\gamma$ with the isomorphism $R/H_\gamma \to R_\gamma$ given by (a). Together, these maps ϕ_γ induce a ring map $\phi : R \to \Pi R_\gamma$, and since each ϕ_γ is surjective we see that the image $P = \phi(R)$ is a subdirect product of the R_γ. In view of (b), we have $\ker \phi = 0$, whence $R \cong P$.

Setting $K_\gamma = \phi(H_\gamma)$ for all γ, we see from (c) that $K_\gamma \in L^*(P_P)$. For any λ, we have $\bigcap_{\gamma \neq \lambda} H_\gamma \subseteq \ker \phi_\gamma$ for all $\gamma \neq \lambda$; hence $E_\lambda = \phi(\bigcap_{\gamma \neq \lambda} H_\gamma)$ must be contained in $P \cap R_\lambda$. In view of (d), we also have $\oplus E_\gamma \in \mathscr{S}(P)$.

Since P is a subdirect product of the R_γ, we must have $p_\lambda(P) = R_\lambda$ for each projection $p_\lambda : \Pi R_\gamma \to R_\lambda$. Now $\ker \phi_\lambda = H_\lambda$, from which we infer that $P \cap (\ker p_\lambda) = K_\lambda$. Inasmuch as $K_\lambda \in L^*(P_P)$ and $\oplus E_\gamma \in \mathscr{S}(P)$, it now follows from 1.28 that $p_\lambda(\oplus E_\gamma) \in \mathscr{S}(R_\lambda)$, that is, $E_\lambda \in \mathscr{S}(R_\lambda)$. Since $E_\lambda \subseteq P \cap R_\lambda$, we thus obtain $P \cap R_\lambda \in \mathscr{S}(R_\lambda)$ for each λ, whence 4.10 says that P must be a right essential product of the R_γ. □

Theorem 4.17 Any right nonsingular ring R is isomorphic to a right essential product of right nonsingular rings S and T such that $\mathrm{soc}(S_S) \leqq_e S_S$ and $\mathrm{soc}(T_T) = 0$.

Proof: Let H denote the *left* annihilator of $\mathrm{soc}(R_R)$ in R, and note that H is a two-sided ideal of R. In view of 1.26, we see that $\mathrm{soc}(H_R) = 0$, and consequently $H \cap \mathrm{soc}(R_R) = 0$. For any nonzero right ideal I of R, either $\mathrm{soc}(I) \neq 0$, in which case $I \cap \mathrm{soc}(R_R) \neq 0$, or else $\mathrm{soc}(I) = 0$, in which case it follows from 1.26 that $I \leqq H$. As a result, we infer that $H \oplus \mathrm{soc}(R_R) \leqq_e R_R$.

Now let K and L denote the respective $\mathscr{S}$-closures of H and $\operatorname{soc}(R_R)$ in R_R, and note that K and L are two-sided ideals of R. Since $Z_r(R) = 0$, it follows from 2.3 that $H_R \leqq_e K_R$ and $\operatorname{soc}(R_R) \leqq_e L_R$, from which we infer that $K \cap L = 0$ and $K \oplus L \leqq_e R_R$. Note also that $K \oplus \operatorname{soc}(R_R) \leqq_e R_R$.

Inasmuch as K and L are two-sided ideals of R which belong to $L^*(R_R)$, 1.28 says that $S = R/K$ and $T = R/L$ are right nonsingular rings. Since $K \cap L = 0$ and $K \oplus L \in \mathscr{S}(R)$, 4.16 now shows that R is isomorphic to a right essential product of S and T.

Since $K \oplus \operatorname{soc}(R_R) \in \mathscr{S}(R)$, 1.28 says that $[K \oplus \operatorname{soc}(R_R)]/K \in \mathscr{S}(S)$. Now $[K \oplus \operatorname{soc}(R_R)]/K$ is isomorphic to $\operatorname{soc}(R_R)$ and so is a semisimple right R-module, hence also a semisimple right S-module. Thus $[K \oplus \operatorname{soc}(R_R)]/K \leqq \operatorname{soc}(S_S)$, whence $\operatorname{soc}(S_S) \leqq_e S_S$. Similarly, since $K \oplus L \in \mathscr{S}(R)$, we use 1.28 again to see that $(K \oplus L)/L \in \mathscr{S}(T)$. Now $(K \oplus L)/L$ is isomorphic to K and so has zero socle as a right R-module, hence also as a right T-module. Thus $[(K \oplus L)/L] \cap [\operatorname{soc}(T_T)] = 0$, from which we conclude that $\operatorname{soc}(T_T) = 0.$ □

Exercises

1. Let $\{R_\gamma \mid \gamma \in X\}$ be a collection of rings. Show that a ring R is isomorphic to a subdirect product of the R_γ if and only if R has two-sided ideals $\{H_\gamma \mid \gamma \in X\}$ such that $R/H_\gamma \cong R_\gamma$ for all γ and $\cap H_\gamma = 0$.
2. Find an example of a subdirect product of two rings which is neither a right essential product nor a left essential product.
3. Let F be a field. Find an example of a right essential product of $\begin{pmatrix} F & 0 \\ F & F \end{pmatrix}$ and $F[x]$ which is not a left essential product.
4. Let $\{R_\gamma\}$ be a collection of rings, and for each γ let E_γ be a two-sided ideal of R_γ which belongs to $\mathscr{S}(R_\gamma)$. If there exists a ring T and isomorphisms $\phi_\gamma : R_\gamma/E_\gamma \to T$ for all γ, show that
$$\{x \in \Pi R_\gamma \mid \phi_\gamma(x_\gamma + E_\gamma) = \phi_\lambda(x_\lambda + E_\lambda) \text{ for all } \gamma,\lambda\}$$
is a right essential product of the R_γ.
5. Set $R_1 = R_2 = R_3 = \mathbf{Z}$, and let R be the set of all triples $(x_1,x_2,x_3) \in R_1 \times R_2 \times R_3$ such that $x_1 \equiv x_2 \pmod 2$ and $x_1 \equiv x_3 \pmod 3$. Show that R is a right essential product of R_1,R_2,R_3, but that R cannot be obtained as in Exercise 4.
6. Set $R_n = \mathbf{Z}/4\mathbf{Z}$ for $n = 1,2,\ldots$, and let R be the subring of ΠR_n generated by 1 and $\oplus(2R_n)$. Show that R is a right essential product of the R_n, but that $R_R \nleqq_e (\Pi R_n)_R$.
7. In 4.14, show that a right ideal I of R belongs to $\mathscr{S}(R)$ if and only if $I[\oplus(R \cap R_\gamma)] \in \mathscr{S}(T)$. Also, show that $I \in \mathscr{S}(R)$ if and only if I contains a right ideal of the form $\oplus I_\gamma$, where each $I_\gamma \in \mathscr{S}(R_\gamma)$.
8. Set $R_n = \mathbf{Z}/4\mathbf{Z}$ for $n = 1,2,\ldots$, and let R be the subring of ΠR_n generated by 1 and $\Pi(2R_n)$. Show that R is a right essential product of the R_n. Show that $\oplus(2R_n) \in \mathscr{S}(\Pi R_n)$, but that $\oplus(2R_n) \notin \mathscr{S}(R)$.

9. Let $Z_r(R) = 0$ and assume that R_R is finite-dimensional. Prove that R is semiprime if and only if R is isomorphic to a right essential product of right nonsingular prime rings.
10. Show that the ring R of Exercise 5 is a finite-dimensional nonsingular semiprime ring. Show that R is not isomorphic to a direct product of prime rings.
11. Let $F_1, F_2, \ldots$ be fields, $R = (\Pi F_n)/(\bigoplus F_n)$. Show that R is a nonsingular semiprime ring which cannot be represented as an essential product of prime rings. (*Hint*: Exercise 3.C.6.)
12. Find an example of a right nonsingular ring R which is not isomorphic to any direct product $S \times T$ such that $\mathrm{soc}(S_S) \leqq_e S_S$ and $\mathrm{soc}(T_T) = 0$.
13. Let A be a simple non-artinian ring, B a simple right A-module, $C = \mathrm{End}_A(B)$. Show that $\begin{pmatrix} A & 0 \\ B & C \end{pmatrix}$ is not isomorphic to any right essential product of rings S,T such that $\mathrm{soc}(S_S) \leqq_e S_S$ and $\mathrm{soc}(T_T) = 0$.
14. Let $\{R_\gamma \mid \gamma \in X\}$ be a collection of rings, and suppose that X is the disjoint union of a collection $\{X_i \mid i \in I\}$ of nonempty subsets. If R_i is a right essential product of $\{R_\gamma | \gamma \in X_i\}$ for all i, and if R is a right essential product of $\{R_i | i \in I\}$, prove that R is also a right essential product of $\{R_\gamma \mid \gamma \in X\}$.
15. Let R be a right essential product of rings $\{R_\gamma \mid \gamma \in X\}$, and assume that X is a disjoint union of nonempty finite subsets X_i. Prove that R is also a right essential product of rings R_i such that each R_i is a right essential product of the rings $\{R_\gamma \mid \gamma \in X_i\}$.
16. If R is a right essential product of right nonsingular rings R_γ, prove that $\mathrm{soc}(R_R) = \bigoplus \mathrm{soc}((R_\gamma)_{R_\gamma})$.
17. If R is a right essential product of rings R_γ, show that R_R is finite-dimensional if and only if each $(R_\gamma)_{R_\gamma}$ is finite-dimensional and all but finitely many $R_\gamma = 0$.
18. If R is a right essential product of right nonsingular rings R_γ, prove that $L^*(R_R) \cong \Pi L^*((R_\gamma)_{R_\gamma})$.
19. If R is a right essential product of rings R_γ, show that R is right noetherian if and only if each R_γ is right noetherian and all but finitely many $R_\gamma = 0$.
20. Let R be a right essential product of two right nonsingular rings R_1 and R_2, and set $E_i = R \cap R_i$ for each i. Prove that a right R-module A is nonsingular if and only if A has a submodule B such that B is a nonsingular right (R/E_1)-module and A/B is a nonsingular right (R/E_2)-module. Prove the corresponding statement for singular modules.
21. Let R be a right essential product of finitely many integral domains $R_1, \ldots, R_n$. If R is right semihereditary, prove that $R = \Pi R_i$.
22. Find an example of a right hereditary ring R such that R is a right essential product of fields $F_1, F_2, \ldots$ but $R \neq \Pi F_n$.
23. Let R be a right essential product of two principal right ideal domains R_1, R_2. If $R \neq R_1 \times R_2$, prove that $\mathrm{r.gl.dim.}(R) = \infty$.

C. Subidealizers

Definition Given a right or left ideal M in a ring T, the *idealizer of M in T* is the largest subring of T which contains M as a two-sided ideal. Thus if M is a right ideal of T, then its idealizer is $\{t \in T \mid tM \subseteq M\}$, while if M is a left ideal of T, then its idealizer is $\{t \in T \mid Mt \subseteq M\}$. For example, if T is the ring of all 2×2 matrices over a ring F, then the idealizer of the right ideal $\begin{pmatrix} 0 & 0 \\ F & F \end{pmatrix}$ is just $\begin{pmatrix} F & 0 \\ F & F \end{pmatrix}$. More generally, the idealizer of a direct summand of T is always a formal triangular matrix ring (Exercise 1).

In general, the idealizer of a right ideal M of T has little in common with T unless some restriction is placed on M. The most useful such restriction is given by the following definition.

Definition A *semimaximal right ideal* of T is any right ideal which is an intersection of finitely many maximal right ideals of T. For example, the semimaximal ideals of $\mathbf{Z}$ are exactly the ideals generated by square-free integers (Exercise 2). An alternate characterization of semimaximal right ideals is given by the following easy lemma.

Lemma 4.18 A right ideal M in a ring T is semimaximal if and only if T/M is a semisimple module.

Proof: If M is semimaximal, then $M = M_1 \cap \cdots \cap M_n$ for suitable maximal right ideals M_i of T. Then T/M is isomorphic to a submodule of the semisimple module $(T/M_1) \oplus \cdots \oplus (T/M_n)$ and hence is semisimple.

Conversely, if T/M is semisimple, then it must be a direct sum of simple modules. Since T/M is also finitely generated, we thus obtain $T/M = S_1 \oplus \cdots \oplus S_n$ for some simple modules S_i. For each j, $\bigoplus_{i \neq j} S_i = M_j/M$ for some maximal right ideal M_j of T, and clearly $M_1 \cap \cdots \cap M_n = M$. Therefore M is semimaximal. □

Proposition 4.19 Let M be a semimaximal right ideal in a ring T, and let R be the idealizer of M in T.

(a) R/M is a semisimple ring.

(b) T/R is a semisimple right R-module.

(c) There is a semimaximal right ideal $M' \geqq M$ in T such that $TM' = T$ and R is also the idealizer of M' in T.

Proof: (a) According to 4.18, T/M is a finitely generated semisimple right T-module, whence $\mathrm{End}_T(T/M)$ is a semisimple ring. Thus it suffices to show that $R/M \cong \mathrm{End}_T(T/M)$.

For any $r \in R$, we have $rM \leqq M$, hence left multiplication by r induces a T-homomorphism $\phi r : T/M \to T/M$. Clearly ϕ is a ring homomorphism of R into $\mathrm{End}_T(T/M)$ with kernel M. Given any $f \in \mathrm{End}_T(T/M)$, we have

$(f\bar{1})M = f(\bar{M}) = 0$; hence $f\bar{1} = \bar{r}$ for some $r \in T$ such that $rM \leqq M$, that is, some $r \in R$. Since f is a right T-homomorphism, we infer that $\phi r = f$. Thus ϕ is surjective; hence it induces a ring isomorphism of R/M onto $\mathrm{End}_T(T/M)$, as desired.

(b) Now $M = M_1 \cap \cdots \cap M_n$ for some maximal right ideals M_i of T, and we set $T_i = \{t \in T \mid tM \leqq M_i\}$ for each i. Note that each T_i is a submodule of T_R which contains R, and that $T_1 \cap \cdots \cap T_n = R$. We claim that each of the right R-modules T/T_i is either 0 or simple.

Suppose that $T/T_i \neq 0$, and consider any nonzero submodule $W/T_i \leqq (T/T_i)_R$. Then W contains an element $w \notin T_i$, whence $wM \nleqq M_i$. Inasmuch as wM and M_i are right ideals of T with M_i maximal, we obtain $wM + M_i = T$. Noting that $wM \leqq wR \leqq W$ and $M_i \leqq T_i \leqq W$, we infer that $W/T_i = T/T_i$. Therefore $(T/T_i)_R$ is either zero or simple, as claimed.

Thus $(T/T_1) \oplus \cdots \oplus (T/T_n)$ is a semisimple right R-module. Since T/R is isomorphic to a submodule of this module, we conclude that $(T/R)_R$ is semisimple.

(c) Inasmuch as $(T/M)_T$ is semisimple, we must have $T/M = (TM/M) \oplus (M'/M)$ for some right ideal $M' \geqq M$ of T. Note that T/M' is isomorphic to a direct summand of the semisimple module T/M, so that M' is a semimaximal right ideal of T. Also, since $M \leqq M'$, we find that $T = TM + M' \leqq TM'$, whence $TM' = T$.

Note that $(T/M')M = (TM + M')/M' = T/M'$. Inasmuch as every submodule of $(T/M')_T$ is a direct summand, it follows that $AM = A$ for all $A \leqq (T/M')_T$. On the other hand, $(T/TM)M = 0$, from which we infer that $\mathrm{Hom}_T(T/TM, T/M') = 0$, and consequently $\mathrm{Hom}_T(M'/M, TM/M) = 0$. Since $T/M = (TM/M) \oplus (M'/M)$, we thus find that M'/M is a fully invariant submodule of T/M. Now left multiplication by elements of R induce endomorphisms of T/M; hence we obtain $RM'/M \leqq M'/M$ and thus $RM' \leqq M'$. Consequently, R is contained in the idealizer of M'. On the other hand, if $t \in T$ and $tM' \leqq M'$, then $t(TM \cap M') \leqq TM \cap M'$ and so $tM \leqq M$. Therefore R equals the idealizer of M' in T.□

Since a number of further properties of idealizers of semimaximal right ideals are consequences only of 4.19(a) and (c), we isolate these conditions as follows.

Definition Let M be a right or left ideal of a ring T. A *subidealizer of M in T* is any subring of T which contains M as a two-sided ideal, i.e., any subring of the idealizer of M which contains M. For example, if S is a subring of the center of T, then $S + M$ is always a subidealizer of M. All subrings of T are subidealizers of 0, but except for this possibility subidealizers of right ideals of T need not also be subidealizers of left ideals of T (Exercise 3).

Definition Let M be a right or left ideal in T. A subidealizer R of M is *tame* provided R/M is a semisimple ring. For example, if M is semimaximal, 4.19 shows that the idealizer of M is a tame subidealizer of M. Also, if S is a semisimple subring of the center of T, then $S + M$ is a tame subidealizer of M.

Definition A *generative right ideal* of T is any right ideal M such that $TM = T$. For instance, every nonzero right ideal of a simple ring is generative. Also, 4.19 shows that the idealizer of any semimaximal right ideal of T can be expressed as the idealizer of a generative semimaximal right ideal. We note that any generative right ideal of T is a generator in Mod-T, although the converse is not true in general (Exercise 4). (See also Exercise 5.)

Proposition 4.20 Let M be a generative right ideal of T, and let R be any subidealizer of M in T. Then T_R and ${}_RM$ are finitely generated projective modules, and the multiplication map $T \otimes_R T \to T$ is an isomorphism.

Proof: The assumption $TM = T$ means that $t_1m_1 + \cdots + t_km_k = 1$ for some $t_i \in T$, $m_i \in M$. Now $m_iT \leqq M \leqq R$ for each i; hence left multiplication by m_i defines a homomorphism $f_i\colon T_R \to R_R$. Observing that $t_1(f_1x) + \cdots + t_k(f_kx) = x$ for all $x \in T$, we see from the Dual Basis Lemma that T_R is finitely generated and projective. Similarly, $Mt_i \leqq M \leqq R$ for each i; hence right multiplication by t_i defines a homomorphism $g_i\colon {}_RM \to {}_RR$. Since $[g_1(x)]m_1 + \cdots + [g_k(x)]m_k = x$ for all $x \in M$, we find that ${}_RM$ is finitely generated and projective.

Given any $x,y \in T$, we have $m_iy \in M \leqq R$ for all i, whence

$$\begin{aligned} x \otimes y &= (xt_1m_1 + \cdots + xt_km_k) \otimes y \\ &= xt_1 \otimes m_1y + \cdots + xt_k \otimes m_ky \\ &= (xt_1m_1y + \cdots + xt_km_ky) \otimes 1 = xy \otimes 1 \end{aligned}$$

in $T \otimes_R T$. Thus the natural map $T \otimes_R R \to T \otimes_R T$ must be surjective. Since the composition $T \otimes_R R \to T \otimes_R T \to T$ is an isomorphism, we conclude that $T \otimes_R T \to T$ must be an isomorphism.□

Corollary 4.21 Let M be a generative right ideal of T, and let R be any subidealizer of M in T.

(a) For any A_T, the natural maps $A \otimes_R T \to A$ and $A \to A \otimes_R T$ are T-module isomorphisms.

(b) For any ${}_TA$, the natural maps $T \otimes_R A \to A$ and $A \to T \otimes_R A$ are T-module isomorphisms.

(c) If A and B are right (left) T-modules, then $\mathrm{Hom}_R(A,B) = \mathrm{Hom}_T(A,B)$.

Proof: According to 4.20, the multiplication map $T \otimes_R T \to T$ is an isomorphism.

(a) The map $A \otimes_R T \to A$ is the composition of the following isomorphisms:

$$A \otimes_R T \to A \otimes_T T \otimes_R T \to A \otimes_T T \to A$$

Inasmuch as the composition $A \to A \otimes_R T \to A$ is the identity map on A, it follows that the map $A \to A \otimes_R T$ is an isomorphism also.

(b) is proved in the same manner.

(c) Given A_T, B_T, and any map $f \in \mathrm{Hom}_R(A,B)$, we have a commutative diagram as follows:

$$\begin{array}{ccc} A & \xrightarrow{f} & B \\ \downarrow & & \downarrow \\ A \otimes_R T & \xrightarrow{f \otimes 1} & B \otimes_R T \end{array}$$

The map $f \otimes 1$ is a T-module homomorphism, and by (a) the vertical maps are T-module isomorphisms; hence we conclude that $f \in \mathrm{Hom}_T(A,B)$. Therefore $\mathrm{Hom}_R(A,B) = \mathrm{Hom}_T(A,B)$. Likewise for left T-modules.□

If R is a subidealizer of a generative right ideal of T, then 4.21 says in particular that the natural maps $R \otimes_R T \to T \otimes_R T$ and $T \otimes_R R \to T \otimes_R T$ are isomorphisms, from which it follows that $(T/R) \otimes_R T = 0$ and $T \otimes_R (T/R) = 0$. Also, 4.21(c) can be extended to the derived functors Ext and Tor, as in Exercise 6.

Returning to the situation in 4.20, we have T_R finitely generated and projective, but no information about ${}_RT$. In general, ${}_RT$ need not be finitely generated or projective, or even flat. (See Exercises 7 to 9.) However, if we assume that M is semimaximal in T and that R is the idealizer of M in T, then we can at least get ${}_RT$ to be flat, as the next theorem shows.

Theorem 4.22 Let M be a generative right ideal of T, and let R be any tame subidealizer of M in T. Then ${}_RT$ is flat if and only if M is a semimaximal right ideal of T and R is the idealizer of M in T.

Proof: First assume that M is semimaximal and that R is the idealizer of M. Note from 4.19 that T/R is a semisimple right R-module. If $x \in T$ and $J = \{r \in R \mid xr \in R\}$, then $R/J \cong (xR + R)/R \leqq T/R$; hence R/J is isomorphic to a direct summand of $(T/R)_R$. As noted above, $(T/R) \otimes_R T = 0$, from which we obtain $(R/J) \otimes_R T = 0$, and thus $JT = T$. According to 3.9, it now follows that ${}_RT$ is flat.

Conversely, assume that ${}_RT$ is flat, and let S denote the idealizer of M in

T. Since $SM \leqq M \leqq R$, we see that S/R is a right module over R/M. Now R/M is a semisimple ring (by hypothesis), whence S/R is a projective right (R/M)-module. Thus if $S/R \neq 0$, there must be a nonzero homomorphism $f: (S/R)_{R/M} \to (R/M)_{R/M}$. Now $f(S/R)$ is a nonzero right ideal of R/M; hence $[f(S/R)]T$ is a nonzero submodule of $(T/M)_T$. Inasmuch as f induces an epimorphism of $(S/R)\otimes_R T$ onto $[f(S/R)]T$, we see that $(S/R)\otimes_R T \neq 0$. On the other hand, we have shown above that $(T/R)\otimes_R T = 0$; hence the map $(S/R)\otimes_R T \to (T/R)\otimes_R T$ is not a monomorphism, which contradicts the flatness of ${}_RT$. Therefore $S/R = 0$, so that R is the idealizer of M in T.

Since R/M is a semisimple ring, we must have $M = M_1 \cap \cdots \cap M_n$ for some maximal right ideals M_i of R. For each i, we claim that $(T/M_iT)_T$ is either zero or simple.

Thus suppose that $T/M_iT \neq 0$, and let J/M_iT be any proper right T-submodule of T/M_iT. Since $1 \notin J$, $J \cap R$ is a proper right ideal of R which contains M_i, whence $J \cap R = M_i$. Then we have a monomorphism $R/M_i \to T/J$; hence by tensoring with the flat module ${}_RT$, we obtain another monomorphism $(R/M_i) \otimes_R T \to (T/J) \otimes_R T$. According to 4.21, the natural map $(T/J)\otimes_R T \to T/J$ is an isomorphism, from which we see that the map $T/M_iT \to T/J$ is a monomorphism. Thus $M_iT = J$, so that $J/M_iT = 0$. Therefore $(T/M_iT)_T$ is either zero or simple, as claimed.

Consequently, $(T/M_1T) \oplus \cdots \oplus (T/M_nT)$ is a semisimple right T-module. Using the flatness of ${}_RT$ again, we see that the monomorphism $R/M \to (R/M_1) \oplus \cdots \oplus (R/M_n)$ induces a monomorphism of T/M into $(T/M_1T) \oplus \cdots \oplus (T/M_nT)$. Therefore T/M is a semisimple right T-module, whence M is a semimaximal right ideal of T.□

Proposition 4.23 Let $Z_r(T) = 0$ and $M \in \mathscr{S}(T)$, and let R be any subidealizer of M in T.

(a) $Z_r(R) = 0$ and $S^\circ R = S^\circ T$.

(b) $\mathscr{S}(T) = \{J \leqq T_T \mid J \cap R \in \mathscr{S}(R)\}$.

(c) $\mathscr{S}(R) = \{K \leqq R_R \mid KT \in \mathscr{S}(T)\}$.

(d) $Z_R(A) = Z_T(A)$ for all A_T.

Proof: We first show that $R_R \leqq_e (S^\circ T)_R$. Given any nonzero $x \in S^\circ T$, we must have $xM \neq 0$, because $M \in \mathscr{S}(T)$ and $(S^\circ T)_T$ is nonsingular. Now xM is a nonzero right T-submodule of $S^\circ T$, and $M_T \leqq_e T_T \leqq_e (S^\circ T)_T$; hence we obtain $xM \cap M \neq 0$, and consequently $xR \cap R \neq 0$. Thus $R_R \leqq_e (S^\circ T)_R$, as claimed.

Inasmuch as $S^\circ T$ is a regular right self-injective ring, (a) now follows from 2.11. According to 2.27, T is a right quotient ring of R; hence (b) to (d) all follow directly from 2.32.□

Certain parts of 4.23 may fail if $Z_r(T) \neq 0$, as shown in Exercises 10 and 11.

We now turn to deriving some conditions under which a subidealizer is hereditary. We proceed by first considering the case of the idealizer of a semimaximal right ideal.

Theorem 4.24 Let M be a semimaximal right ideal of T, and let R be the idealizer of M in T. Then R is right hereditary if and only if T is right hereditary.

Proof: In view of 4.19, there is no loss of generality in assuming that $TM = T$, so that the results of 4.20 and 4.21 are applicable.

First assume that R is right hereditary, and consider any right ideal J of T. Since T_R is projective by 4.20, J_R must be projective, from which it follows that $(J \otimes_R T)_T$ is projective. According to 4.21, $J \otimes_R T \cong J$, whence J_T is projective. Thus T is right hereditary.

Now assume that T is right hereditary, and let K be any right ideal of R. Then $(KT)_T$ and T_R are both projective, from which it follows that $(KT)_R$ is projective.

There is an obvious epimorphism f of some direct sum $\oplus A_\alpha$ of copies of $(T/R)_R$ onto $(KT/K)_R$. According to 4.19, $(T/R)_R$ is semisimple; hence so is $\oplus A_\alpha$, from which we see that $\oplus A_\alpha \cong (KT/K) \oplus (\ker f)$. Since each $A_\alpha \cong (T/R)_R$, there exists a short exact sequence $0 \to F \to G \to \oplus A_\alpha \to 0$ of right R-modules such that F_R and G_T are both free. We must also have $\ker f \cong H/L$ for some free H_R and some $L \leqq H$, from which we obtain another short exact sequence $0 \to K \oplus L \to KT \oplus H \to \oplus A_\alpha \to 0$. We have seen that $(KT)_R$ is projective, and G_R is projective because T_R is projective; hence Schanuel's Lemma says that $G \oplus (K \oplus L) \cong (KT \oplus H) \oplus F$. Inasmuch as $KT \oplus H \oplus F$ is a projective right R-module, we conclude that K_R is projective.

Therefore R is right hereditary.□

Corollary 4.25 Let $Z_r(T) = 0$, let M be a generative essential right ideal of T, and let R be any tame subidealizer of M in T. Then R is right hereditary if and only if

(a) M is a semimaximal right ideal of T.
(b) R is the idealizer of M in T.
(c) T is right hereditary.

Proof: If (a) to (c) hold, then R is right hereditary by 4.24.

Now assume that R is right hereditary. According to 4.23, we have $Z_r(R) = 0$ and $S^\circ R = S^\circ T$, so that T is a subring of $S^\circ R$. Since all finitely generated right ideals of R are projective and thus finitely presented, R is right coherent, whence 3.7 says that ${}_R(S^\circ R)$ is flat. Inasmuch as R is right hereditary and ${}_RT \leqq {}_R(S^\circ R)$, we conclude that ${}_RT$ is flat; hence (a) and (b) follow from 4.22. Finally, (c) follows from 4.24.□

We conclude this section by deriving some conditions under which a subidealizer is noetherian.

Theorem 4.26 Let M be a generative semimaximal right ideal of T, let S be the idealizer of M in T, and let R be any tame subidealizer of M in T. Then R is right noetherian if and only if T is right noetherian and S_R is finitely generated.

Proof: Note from 4.20 that T_R and T_S are both finitely generated.

If R is right noetherian, then T_R must be noetherian, from which it follows that S_R is finitely generated and that T_T is noetherian.

Conversely, assume that T is right noetherian and that S_R is finitely generated. According to 4.19, T/S is a semisimple right S-module. Since T_S is finitely generated, we see that $(T/S)_S$ is noetherian. Also, 4.19 says that S/M is a semisimple ring, whence $(S/M)_S$ is noetherian, and consequently $(T/M)_S$ is noetherian.

If J is any right ideal of R, then JT and JM are right ideals of T, whence $(JT)_T$ and $(JM)_T$ are finitely generated. Since T_R is finitely generated, it follows that $(JT)_R$ and $(JM)_R$ are finitely generated as well. Inasmuch as $(JT)_T$ is finitely generated, we must have $JT = x_1T + \cdots + x_nT$ for suitable $x_i \in J$. Then each $x_iM \leqq JM$, whence JT/JM is an epimorphic image of a direct sum of n copies of T/M. Since $(T/M)_S$ is noetherian, it follows that $(JT/JM)_S$ is noetherian, whence $(JS/JM)_S$ must be finitely generated. Also, S_R is finitely generated; hence we see that $(JS/JM)_R$ is finitely generated.

Inasmuch as $SM \leqq M \leqq R$, JS/JM is a right module over the semisimple ring R/M, from which we infer that $(JS/JM)_R$ is noetherian. Consequently, $(J/JM)_R$ is finitely generated. Since $(JM)_R$ is finitely generated as well, we conclude that J_R is finitely generated. Therefore R is right noetherian.□

There are results similar to 4.26 for other chain conditions, as in Exercises 12 and 13.

Exercises

1. If $e = e^2 \in T$, show that the idealizer of eT in T is isomorphic to
$$\begin{pmatrix} (1-e)T(1-e) & 0 \\ eT(1-e) & eTe \end{pmatrix}.$$
2. Show that an integer n generates a semimaximal ideal of $\mathbf{Z}$ if and only if n is not divisible by the square of any prime.

3. Let T be a simple non-artinian ring, M a proper essential right ideal of T, R the idealizer of M in T. Show that R is not a subidealizer of any nonzero left ideal of T.
4. Find an example of a ring T with a right ideal M such that M_T is a generator in Mod-T but M is not a generative right ideal of T.
5. If T is a regular ring, prove that a right ideal M of T is generative if and only if M_T is a generator in Mod-T.
6. Let M be a generative right ideal of T, R any subidealizer of M in T. Prove that $\mathrm{Tor}_n^R(A,B) \cong \mathrm{Tor}_n^T(A,B)$ for all A_T, all ${}_TB$, and all $n > 0$. Prove that $\mathrm{Ext}_R^n(A,C) \cong \mathrm{Ext}_T^n(A,C)$ for all right (left) T-modules A,C and all $n > 0$.
7. Let $K \subseteq L$ be fields with $[L : K] = \infty$, and set $T = \begin{pmatrix} L & L \\ L & L \end{pmatrix}$, $M = \begin{pmatrix} 0 & 0 \\ L & L \end{pmatrix}$, $R = \begin{pmatrix} K & 0 \\ L & L \end{pmatrix}$. Show that M is a generative semimaximal right ideal of T and that R is a tame subidealizer of M in T, but that ${}_RT$ is neither flat nor finitely generated.
8. Let V be an infinite-dimensional vector space over a division ring D, $T = \mathrm{End}_D(V)$, $e \in T$ a projection onto some one-dimensional subspace of V. If R is the idealizer of $(1 - e)T$ in T, show that ${}_RT$ is not finitely generated.
9. Make $T = \begin{pmatrix} \mathbf{Z} & \mathbf{Z}/2\mathbf{Z} \\ \mathbf{Z}/2\mathbf{Z} & \mathbf{Z} \end{pmatrix}$ into a ring by using the usual matrix operations together with the zero multiplication in $\mathbf{Z}/2\mathbf{Z}$. If R is the idealizer of $\begin{pmatrix} 0 & 0 \\ \mathbf{Z}/2\mathbf{Z} & \mathbf{Z} \end{pmatrix}$ in T, show that ${}_RT$ is not flat.
10. With notation as in Exercise 4.B.8, set $T = \Pi R_n$ and $M = \bigoplus(2R_n)$. Show that $M \in \mathscr{S}(T)$ and that R is a subidealizer of M in T, but that $M \notin \mathscr{S}(R)$.
11. With notation as in Exercise 4.B.6, set $T = \Pi R_n$ and $M = \bigoplus(2R_n)$. Show that $M \in \mathscr{S}(T)$ and that R is a subidealizer of M in T, but that T is not a right quotient ring of R.
12. Let S be a generative semimaximal right ideal of T, S the idealizer of M in T, R any tame subidealizer of M in T. Prove that R is right artinian if and only if T is right artinian and S_R is finitely generated.
13. Let $Z_r(T) = 0$, $M \in \mathscr{S}(T)$, R any subidealizer of M in T. Show that R_R is finite-dimensional if and only if T_T is finite-dimensional. Also, show that R is right Goldie if and only if T is right Goldie.
14. Let $Z_r(T) = 0$, $M \in \mathscr{S}(T)$, R any subidealizer of M in T. Show that R is prime (semiprime) if and only if T is prime (semiprime).
15. Let R be the idealizer of a semimaximal right ideal of T. Prove that any simple right T-module is either simple as an R-module or else has an R-module composition series of length 2.
16. Let M be a semimaximal right ideal of T, and let R be the idealizer of M in T. Prove that there is a bijection between the right ideals of R/M and the submodules of $(T/M)_T$.
17. Let M be a generative semimaximal right ideal of T, and let R be the idealizer of M in T. Prove that there is a bijection between the submodules of $(T/M)_T$ and the submodules of $(T/R)_R$.
18. Let $Z_r(T) = 0$, M an essential semimaximal right ideal of T, R the idealizer of

M in T. If A is any nonsingular right R-module, prove that $A \otimes_R T$ is a nonsingular right T-module and that the natural map $A \to A \otimes_R T$ is injective.

19. If M is a generative right ideal of T and R is a tame subidealizer of M in T, prove that

$$\text{r.gl.dim.}(T) \leqq \text{r.gl.dim.}(R) \leqq 1 + \text{r.gl.dim.}(T)$$

20. If M is a generative right ideal of T and R is a tame subidealizer of M in T, prove that

$$\text{l.gl.dim.}(T) \leqq \text{l.gl.dim.}(R) \leqq 1 + \text{l.gl.dim.}(T)$$

21. Let M be a generative right ideal of T such that $(T/M)_T$ is flat, and let R be any tame subidealizer of M in T. If T is left semihereditary, prove that R is left semihereditary. [*Hint*: Show that $(R/M)_R$ is flat, and obtain $J \cap M \cong M \otimes_T TJ$ for every left ideal J of R.]

5

Rings Whose Nonsingular Modules Are Projective

The main purpose of this chapter is to derive necessary and sufficient conditions on a right nonsingular ring R under which all nonsingular right R-modules are projective, and to describe the structure of such rings. Along the way, we characterize the following related conditions: (a) All nonsingular right R-modules are flat; (b) Every finitely generated nonsingular right R-module can be embedded in a free right R-module; (c) All finitely generated nonsingular right R-modules are projective. These results are obtained in Section C. In order to develop them, we must first study rings for which all flat right modules are projective (Section A) and rings for which all direct products of projective right modules are projective (Section B).

A. Perfect Rings

A ring R for which all flat right R-modules are projective is known as a "right perfect ring," and the purpose of this section is to derive characterizations of right perfect rings. We begin with some alternate descriptions of flat modules, for which we need the following easy lemma.

Lemma 5.1 Let $0 \to K \to P \to A \to 0$ be an exact sequence of right R-modules, and let $0 \to L \to Q \to B \to 0$ be an exact sequence of left R-modules. If Q is flat and the map $A \otimes_R L \to A \otimes_R Q$ is injective, then the map $K \otimes_R B \to P \otimes_R B$ is injective.

Proof: Since Q is flat, the map $K \otimes_R Q \to P \otimes_R Q$ is injective; hence we obtain a commutative diagram with exact rows and columns as follows:

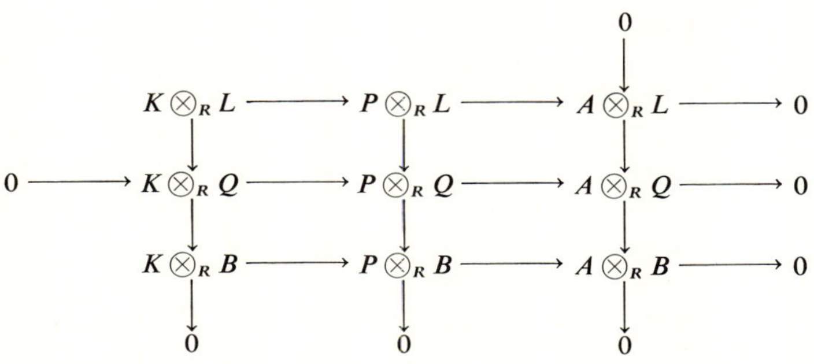

An easy diagram chase now shows that the map $K \otimes_R B \to P \otimes_R B$ is injective. □

Theorem 5.2 Let $0 \to K \xrightarrow{f} P \to A \to 0$ be a short exact sequence of right R-modules. If P is projective, then the following conditions are equivalent:

(a) A is flat.

(b) For all ${}_RB$, the map $f \otimes 1 : K \otimes_R B \to P \otimes_R B$ is injective.

(c) For any $x \in K$, there is a homomorphism $g : P \to K$ such that $gfx = x$.

(d) If H is any finitely generated submodule of K, then there is a homomorphism $g: P \to K$ such that $(1 - gf)(H) = 0$.

Proof: (a) ⇒ (b): Choose an exact sequence $0 \to L \to Q \to B \to 0$ of left R-modules with Q free. Then Q is flat, and the map $A \otimes_R L \to A \otimes_R Q$ is injective because A is flat; hence 5.1 says that $f \otimes 1$ is injective.

(b) ⇒ (a): Given any short exact sequence $0 \to D \to C \to B \to 0$ of left R-modules, we see from (b) that the map $K \otimes_R B \to P \otimes_R B$ is injective; hence it follows from the left-right symmetric version of 5.1 that the map $A \otimes_R D \to A \otimes_R C$ is injective. Therefore A is flat.

(b) ⇒ (c): There exists a free module F and maps $j: P \to F$ and $k: F \to P$ such that $kj = 1_P$. Let Y be a basis for F, and write $jfx = y_1 r_1 + \cdots + y_n r_n$ for suitable $y_i \in Y$, $r_i \in R$. If $I = Rr_1 + \cdots + Rr_n$, then the map $f \otimes 1$: $K \otimes_R (R/I) \to P \otimes_R (R/I)$ is injective by (b). Inasmuch as $(k \otimes 1)(j \otimes 1) = kj \otimes 1$ is the identity map on $P \otimes_R (R/I)$, we see that $j \otimes 1$ is injective, whence $jf \otimes 1$ must be injective. Using the natural isomorphisms $K \otimes_R (R/I) \cong K/KI$ and $F \otimes_R (R/I) \cong F/FI$, we infer from this that $(jf)^{-1}(FI) = KI$.

By definition of I, $jfx \in FI$, whence $x \in KI$. Thus $x = x_1 r_1 + \cdots + x_n r_n$ for some $x_i \in K$. Define a map $h : F \to K$ by setting $hy_i = x_i$ for all i and $hy = 0$ for all other $y \in Y$. Then $hjfx = x$; hence $g = hj$ is a map from P into K such that $gfx = x$.

(c) ⇒ (d): The module H must be generated by elements $x_1, \ldots, x_n \in K$, and it suffices to find a map $g : P \to K$ such that $gfx_i = x_i$ for all i. The case $n = 1$ is (c). Now let $n > 1$, and assume we have a map $h : P \to K$ such that $hfx_i = x_i$ for $i = 1, \ldots, n - 1$. According to (c), there also exists a map $k : P \to K$ such that $kf(x_n - hfx_n) = x_n - hfx_n$. Inasmuch as $x_i - hfx_i = 0$ for $i = 1, \ldots, n - 1$, we thus obtain $kf(x_i - hfx_i) = x_i - hfx_i$ for $i = 1, \ldots, n$. Consequently, $g = k - kfh + h$ is the required map of P into K such that $gfx_i = x_i$ for all i.

(d) $\Rightarrow$ (b): Consider any element $\gamma = x_1 \otimes b_1 + \cdots + x_n \otimes b_n$ in the kernel of $f \otimes 1$. In view of (d), there is a map $g : P \to K$ such that $gfx_i = x_i$ for all i, whence $\gamma = (g \otimes 1)(f \otimes 1)(\gamma) = 0$. Therefore $f \otimes 1$ is injective. □

Theorem 5.2 can be used to show that flatness is preserved by Morita-equivalences (Exercise 1). Thus, for example, any ring Morita-equivalent to a regular ring must be regular. (See also Exercise 2.)

Corollary 5.3 Every finitely presented flat module is projective.

Proof: Given any finitely presented module A, there exists a short exact sequence $0 \to K \xrightarrow{f} F \to A \to 0$ such that F is a finitely generated free module and K is a finitely generated module. If A is also flat, then by 5.2 there is a map $g : F \to K$ such that $(1 - gf)(K) = 0$, whence the sequence splits and A is projective. □

If R is nonsingular, then 5.3 also works for essentially finitely related flat modules (Exercise 3).

Lemma 5.4 Let F be a free right R-module with basis $x_1, x_2, \ldots$. Let $a_1, a_2, \ldots \in R$, and let K be the submodule of F generated by $x_1 - x_2a_1$, $x_2 - x_3a_2, \ldots$.

(a) F/K is flat.

(b) If F/K is projective, then there exists an integer $k > 1$ such that $Ra_ka_{k-1} \cdots a_1 = Ra_{k-1}a_{k-2} \cdots a_1$.

Proof: (a) Given any $x \in K$, there must exist a positive integer n such that x lies in the submodule of K generated by $x_1 - x_2a_1, \ldots, x_n - x_{n+1}a_n$. Define a map $f : F \to K$ by setting $fx_i = x_i - x_{n+1}a_na_{n-1} \cdots a_i$ for $i = 1, \ldots, n$ and $fx_i = 0$ for $i > n$. We check that $f(x_i - x_{i+1}a_i) = x_i - x_{i+1}a_i$ for all $i = 1, \ldots, n$, from which we obtain $fx = x$. Now 5.2 shows that F/K is flat.

(b) We first claim that K is free with basis $x_1 - x_2a_1, x_2 - x_3a_2, \ldots$. Thus suppose that

$$(x_1 - x_2a_1)r_1 + \cdots + (x_n - x_{n+1}a_n)r_n = 0$$

for some n and some $r_i \in R$. As a result, we compute that

$$x_1r_1 + x_2(r_2 - a_1r_1) + \cdots + x_n(r_n - a_{n-1}r_{n-1}) - x_{n+1}a_nr_n = 0$$

Consequently, $r_1 = r_2 - a_1r_1 = \cdots = r_n - a_{n-1}r_{n-1} = 0$, whence $r_1 = \cdots = r_n = 0$. Therefore $x_1 - x_2a_1, x_2 - x_3a_2, \ldots$ is indeed a basis for K; hence there is an isomorphism $f : K \to F$, where $f(x_k - x_{k+1}a_k) = x_k$.

Since F/K is projective, K is a direct summand of F, hence f extends to a map $g : F \to F$ such that $g(x_k - x_{k+1}a_k) = x_k$ for all k. For each $n = 1,2,\ldots$, there exist elements $c_{1n},c_{2n},\ldots$ in R such that $c_{kn} = 0$ for all but finitely many k and $g(x_n) = \sum x_kc_{kn}$. Then

$$x_n = g(x_n - x_{n+1}a_n) = \sum x_k(c_{kn} - c_{k,n+1}a_n)$$

hence we obtain $c_{nn} - c_{n,n+1}a_n = 1$ and $c_{kn} = c_{k,n+1}a_n$ for all $k \neq n$.

Now $c_{k1} = 0$ for some $k > 1$, and we compute that

$$\begin{aligned} c_{kk}a_{k-1}a_{k-2} \cdots a_1 &= c_{k,k-1}a_{k-2}a_{k-3} \cdots a_1 \\ &= \cdots = c_{k2}a_1 = c_{k1} = 0 \end{aligned}$$

As a result,

$$\begin{aligned} a_{k-1}a_{k-2} \cdots a_1 &= (1 - c_{kk})a_{k-1}a_{k-2} \cdots a_1 \\ &= -c_{k,k+1}a_ka_{k-1} \cdots a_1 \end{aligned}$$

from which we conclude that $Ra_ka_{k-1} \cdots a_1 = Ra_{k-1}a_{k-2} \cdots a_1$, as desired. □

The Jacobson radical of a perfect ring turns out to posess a certain weak nilpotence property, as follows.

Definition A two-sided ideal J of R is said to be *right T-nilpotent* if for any sequence $x_1,x_2,\ldots$ of elements of J, there is a positive integer n such that $x_nx_{n-1} \cdots x_1 = 0$. (Similarly, J is *left T-nilpotent* if for any sequence $x_1,x_2,\ldots$ in J, we have $x_1x_2 \cdots x_n = 0$ for some n.) In particular, this must hold for any constant sequence $x,x,\ldots$ (where $x \in J$), in which case the definition requires that $x^n = 0$ for some n. Thus every right T-nilpotent ideal is nil. (The converse fails, as shown by Exercise 4.) It is clear that every nilpotent ideal is right (and left) T-nilpotent, but Exercise 5 shows that not every T-nilpotent ideal is nilpotent. Finally, we note that right T-nilpotent ideals are not always left T-nilpotent (Exercise 6).

Lemma 5.5 Let J be a right T-nilpotent two-sided ideal of R. If A is any right R-module such that $AJ = A$, then $A = 0$.

Proof: If $A \neq 0$, then since $AJ = A$ we must have $Ax_1 \neq 0$ for some

$x_1 \in J$. Then $AJx_1 \neq 0$, whence $Ax_2x_1 \neq 0$ for some $x_2 \in J$. Continuing in this manner, we obtain a sequence $x_1, x_2, \ldots$ of elements of J such that $Ax_nx_{n-1} \cdots x_1 \neq 0$ for all n. But then $x_nx_{n-1} \cdots x_1 \neq 0$ for all n, which contradicts the right T-nilpotence of J.□

Actually, the condition "$AJ = A$ implies $A = 0$" is equivalent to the condition that J is right T-nilpotent. (See Exercise 7.)

Definition Let J be a two-sided ideal of R. We say that *idempotents can be lifted mod J* provided each idempotent in the ring R/J is the image (under the natural map $R \to R/J$) of a suitable idempotent in R. For example, since the only idempotents in the ring $\mathbf{Z}/2\mathbf{Z}$ are 0 and 1, we see that idempotents can be lifted mod $2\mathbf{Z}$. On the other hand, since $\mathbf{Z}/6\mathbf{Z}$ has four distinct idempotents while $\mathbf{Z}$ has only two, idempotents cannot be lifted mod $6\mathbf{Z}$. Also, if Q is the endomorphism ring of any injective module, then 2.18 shows that idempotents can be lifted mod $J(Q)$.

Lemma 5.6 If J is a nil two-sided ideal of R, then idempotents can be lifted mod J.

Proof: We must show that any idempotent $a + J$ in R/J lifts to an idempotent $e \in R$; that is, given $a \in R$ such that $a^2 - a \in J$, we need $e \in R$ such that $e^2 - e = 0$ and $e - a \in J$. Let T denote the subring of R generated by 1 and a, and note that T is commutative. Now $J \cap T$ is a nil ideal of T, $a^2 - a \in J \cap T$, and it suffices to find an element $e \in T$ such that $e^2 - e = 0$ and $e - a \in J \cap T$. Thus we may assume, without loss of generality, that R is commutative.

Since $a^2 - a$ belongs to the nil ideal J, we must have $(a^2 - a)^n = 0$ for some positive integer n. Setting $I = (a^2 - a)R$, we use the commutativity of R to see that $I^n = 0$ as well. It obviously suffices to find an idempotent $e \in R$ such that $e - a \in I$; hence there is no loss of generality in assuming that $J^n = 0$.

We proceed by induction on n. If $n = 1$, then $J = 0$ and $a^2 - a = 0$; hence there is nothing to prove. Next let $n = 2$, and set $z = a^2 - a$, $e = a + z(1 - 2a)$. Inasmuch as $z \in J$, we have $z^2 = 0$ and $e - a \in J$. It follows that $e^2 = a^2 + 2az(1 - 2a)$, whence

$$\begin{aligned} e^2 - e &= a^2 - a + 2az(1 - 2a) - z(1 - 2a) \\ &= z[1 + 2a(1 - 2a) - (1 - 2a)] \\ &= 4z(a - a^2) = -4z^2 = 0 \end{aligned}$$

Thus e is the required idempotent in R such that $e - a \in J$.

Finally, let $n > 2$ and assume that we can lift idempotents in commutative rings mod ideals whose $(n - 1)$st power is zero. Since $n > 2$, we have $(J^2)^{n-1} = 0$; hence we can lift idempotents mod J^2. By the case $n = 2$, we can also lift idempotents from R/J back to R/J^2, from which we conclude that we can lift idempotents mod J.□

Theorem 5.7 The following conditions are equivalent:

(a) All flat *right* R-modules are projective.

(b) R has DCC on principal *left* ideals.

(c) $J(R)$ is *right* T-nilpotent, and $R/J(R)$ is a semisimple ring.

Proof: (a) ⇒ (b): If not, then R has a strictly decreasing sequence $I_1 > I_2 > \cdots$ of principal left ideals, and we observe that there exist elements $a_1, a_2, \ldots \in R$ such that each $I_n = Ra_na_{n-1} \cdots a_1$. Defining F and K as in 5.4, we see from 5.4 that F/K is flat, whence (a) says that F/K is projective. But then 5.4 says that $I_k = I_{k-1}$ for some $k > 1$, which is a contradiction.

(b) ⇒ (c): Given elements $x_1, x_2, \ldots \in J(R)$, (b) says that the sequence $Rx_1 \geqq Rx_2x_1 \geqq \cdots$ of principal left ideals of R must terminate, whence $Rx_nx_{n-1} \cdots x_1 = Rx_{n+1}x_n \cdots x_1$ for some n. Consequently, $x_nx_{n-1} \cdots x_1 = rx_{n+1}x_n \cdots x_1$ for some $r \in R$; hence $(1 - rx_{n+1})x_nx_{n-1} \cdots x_1 = 0$. Since $x_{n+1} \in J(R)$, $1 - rx_{n+1}$ is invertible in R, so that $x_nx_{n-1} \cdots x_1 = 0$. Thus $J(R)$ is right T-nilpotent.

Observing that $R/J(R)$ inherits the DCC on principal left ideals, we see that to prove $R/J(R)$ semisimple it suffices to consider the case when $J(R) = 0$.

Note that any left ideal of R which is minimal among the nonzero principal left ideals of R must be a simple left ideal. As a result, every nonzero left ideal of R contains a simple left ideal. We also claim that any simple left ideal K of R is a direct summand of ${}_RR$. Inasmuch as $J(R) = 0$, R must have a maximal left ideal M such that $K \nleqq M$, whence $K \cap M < K$. The simplicity of K forces $K \cap M = 0$, and then we conclude from the maximality of M that $K \oplus M = R$.

Choose a simple left ideal A_1 of R, and get ${}_RR = A_1 \oplus B_1$ for some B_1. If $B_1 \neq 0$, choose a simple left ideal $A_2 \leqq B_1$. Then A_2 is a direct summand of ${}_RR$ and hence of B_1, so that $B_1 = A_2 \oplus B_2$ for some B_2. Continuing as long as possible in this manner, we obtain simple left ideals $A_1, A_2, \ldots$ and left ideals $B_1 > B_2 > \cdots$ such that ${}_RR = A_1 \oplus \cdots \oplus A_n \oplus B_n$ for each n. Each B_n is a direct summand of ${}_RR$ and so is principal, whence (b) says that the sequence $B_1 > B_2 > \cdots$ must terminate. This happens only when some $B_n = 0$, at which point ${}_RR = A_1 \oplus \cdots \oplus A_n$. Therefore ${}_RR$ is semisimple, and R is a semisimple ring.

(c) ⇒ (a): Set $J = J(R)$, and let A be any flat right R-module. Since R/J is a semisimple ring, A/AJ must be isomorphic to a direct sum of copies of simple right ideals of R/J, each of which is generated by an idempotent in

R/J. We are given that J is right T-nilpotent and hence nil; so by 5.6 each of these idempotents in R/J lifts to an idempotent in R. Thus there is a collection $\{e_\alpha\}$ of idempotents in R such that $\oplus[(e_\alpha R + J)/J] \cong A/AJ$. Since J is a two-sided ideal, we infer that $e_\alpha R \cap J = e_\alpha J$ for each α; hence $A/AJ \cong \oplus(e_\alpha R/e_\alpha J)$.

Now $P = \oplus(e_\alpha R)$ is a projective right R-module such that $A/AJ \cong P/PJ$. The isomorphism $P/PJ \to A/AJ$ must lift to a map $f: P \to A$ such that $f^{-1}(AJ) = PJ$ and $fP + AJ = A$. Set $K = \ker f$, and note that $K \leqq PJ$. Since $fP + AJ = A$, we have $(A/fP)J = A/fP$, whence 5.5 says that $A/fP = 0$. Thus we obtain a short exact sequence $0 \to K \to P \to A \to 0$.

According to 5.2, the flatness of A implies that the map $K \otimes_R (R/J) \to P \otimes_R (R/J)$ is injective, from which we conclude that $K \cap PJ = KJ$. Inasmuch as $K \leqq PJ$, we see that $K = KJ$, and then $K = 0$, by 5.5. Therefore A is isomorphic to P and thus is projective.□

Definition Any ring R which satisfies the conditions of 5.7 is called a *right perfect ring*. For example, all semisimple rings are right (and left) perfect. More generally, if $J(R)$ is nilpotent and $R/J(R)$ is semisimple, then R is right and left perfect. We note that right perfect rings are not always left perfect, as shown by Exercise 8. Exercise 9 shows that any ring Morita-equivalent to a right perfect ring is right perfect. In particular, the ring of all $n \times n$ matrices over a right perfect ring is right perfect.

We conclude this section by showing that all artinian rings are perfect. There is also a partial converse—all noetherian perfect rings are artinian—for which we need the following proposition.

Proposition 5.8 If R is right noetherian, then all nil one-sided ideals of R are nilpotent.

Proof: Because R is right noetherian, we can choose a nilpotent two-sided ideal N of R which is maximal among all nilpotent two-sided ideals of R. It follows from the maximality of N that R/N has no nonzero nilpotent two-sided ideals, that is, R/N is semiprime. It clearly suffices to show that all nil one-sided ideals of R are contained in N. Therefore we may assume, without loss of generality, that R is semiprime, and we must show that all nil one-sided ideals of R are zero.

Suppose on the contrary that R has a nonzero nil one-sided ideal K, and choose a nonzero element $x \in K$. If K is a left ideal, then $Rx \leqq K$ and Rx is a nil left ideal of R. If K is a right ideal, then for any $r \in R$ we have $xr \in K$; hence $(xr)^n = 0$ for some $n > 0$, and consequently $(rx)^{n+1} = r(xr)^n x = 0$. Thus in either case Rx is a nonzero nil left ideal of R.

Using again the noetherian hypothesis, choose a nonzero element $y \in Rx$

whose right annihilator $r(y)$ is maximal among the right annihilators of nonzero elements of Rx. We claim that $yRy = 0$. If $t \in R$ with $ty \neq 0$, then since Rx is nil there must be a positive integer n such that $(ty)^n \neq 0$ and $(ty)^{n+1} = 0$. Thus $(ty)^n$ is a nonzero element of Rx, and we see that $r(y) \leqq r((ty)^n)$; hence it follows from the maximality of $r(y)$ that $r(y) = r((ty)^n)$. Inasmuch as $(ty)^n(ty) = 0$, we thus obtain $y(ty) = 0$. Therefore $yRy = 0$, as claimed. Now $(RyR)^2 = 0$, which contradicts the semiprimeness of R. □

Theorem 5.9 The following conditions are equivalent:

(a) R is right artinian.

(b) R is right noetherian and *right* perfect.

(c) R is right noetherian and *left* perfect.

Proof: Set $J = J(R)$.

(c) ⇒ (a): Since R is right noetherian, each of the right R-modules $R/J, J/J^2, \ldots$ is finitely generated. However, these are also right modules over R/J, which is a semisimple ring (because R is left perfect). Thus each of the right R-modules R/J, $J/J^2, \ldots$ is a finite direct sum of simple modules and hence has a composition series. Consequently, $(R/J^n)_R$ has a composition series for all n, hence each of the rings R/J^n is right artinian. Inasmuch as J is left T-nilpotent and hence nil, 5.8 shows that $J^n = 0$ for some n. Therefore R is right artinian.

(a) ⇒ (c): Since R is right artinian, it must have DCC on principal right ideals, whence R is left perfect. In particular, it follows that R/J is semisimple.

Using DCC on right ideals, we find a positive integer n such that $J^n = J^{n+1}$, that is, $JJ^n = J^n$. Since R is left perfect, J is left T-nilpotent; hence 5.5 says that $J^n = 0$.

Each of the right R-modules $R/J, J/J^2, \ldots$ is also a right module over the semisimple ring R/J and so is a direct sum of simple modules. Since R_R is artinian, each of these direct sums must contain only finitely many simple modules. Thus $(R/J)_R$, $(J/J^2)_R, \ldots, (J^{n-1}/J^n)_R$ are all noetherian, whence $(R/J^n)_R$ is noetherian, that is, R is right noetherian.

(b) ⇒ (a) is proved in the same manner as (c) ⇒ (a).

(a) ⇒ (b): Proceeding as in (a) ⇒ (c), we see that R is right noetherian and that J is nilpotent. In particular, J is right T-nilpotent. Inasmuch as R is left perfect by (c), R/J is a semisimple ring, and consequently R is also right perfect. □

Using 5.9, we obtain a large class of examples of right perfect rings, namely all right or left artinian rings. However, not all perfect rings are artinian, as shown by Exercise 11.

Exercises

1. Prove that any Morita-equivalence Mod-$R \to$ Mod-S carries flat right R-modules to flat right S-modules.
2. Prove that any two Morita-equivalent rings have the same global weak dimension.
3. If $Z_r(R) = 0$, prove that all essentially finitely related flat right R-modules are projective.
4. If R is the direct product of the rings $\mathbf{Z}/2^n\mathbf{Z}$ ($n = 1,2, \ldots$), show that the prime radical of R is nil but not T-nilpotent.
5. Set $R_n = \mathbf{Z}/n\mathbf{Z}$ for $n = 1,2, \ldots$, and let R be the subring of ΠR_n generated by 1 and $\oplus R_n$. Show that $J(R)$ is T-nilpotent but not nilpotent.
6. Let F be a field, and let R be the ring of all upper triangular countably infinite matrices over F with only finitely many nonzero off-diagonal entries. Show that $J(R)$ is right T-nilpotent but not left T-nilpotent.
7. Let J be a two-sided ideal of R. If $AJ \neq A$ for all nonzero A_R, prove that J is right T-nilpotent.
8. Let F,R be as in Exercise 6, and let S be the F-subalgebra of R generated by 1 and $J(R)$. Show that S is right perfect but not left perfect.
9. If R and S are Morita-equivalent rings, show that R is right perfect if and only if S is right perfect.
10. Prove that $\begin{pmatrix} A & 0 \\ B & C \end{pmatrix}$ is right perfect if and only if A and C are both right perfect.
11. Find an example of a right and left perfect ring which is neither right artinian nor left artinian.
12. If $Z_r(R) = 0$ and R_R is finite-dimensional, prove that all flat right R-modules are nonsingular.
13. Let $Z_r(R) = 0$ and assume that R_R is finite-dimensional. Let A be a flat right R-module. If B is any finitely generated submodule of A, prove that there exist elements $a_1, \ldots, a_n \in A$ and maps $f_1, \ldots, f_n \in \mathrm{Hom}_R(B,R)$ such that $a_1(f_1b) + \cdots + a_n(f_nb) = b$ for all $b \in B$.
14. Use 5.5 to show that all right T-nilpotent two-sided ideals of R are contained in $J(R)$.
15. Prove that any factor ring of a right perfect ring is right perfect.
16. Let M be a semimaximal right ideal of T, and let R be the idealizer of M in T. Prove that R is right perfect if and only if T is right perfect.
17. Let J be a two-sided ideal of R such that $J \leqq J(R)$ and idempotents can be lifted mod J. Given orthogonal idempotents $a_1 + J$, $a_2 + J, \ldots$ in R/J, prove that there exist orthogonal idempotents $e_1,e_2, \ldots$ in R such that $e_i + J = a_i + J$ for all i.
18. Prove that R is right perfect if and only if every left R-module has essential socle and R contains no infinite sets of orthogonal idempotents.
19. Let R be right perfect. If P and Q are projective right R-modules such that $P/PJ(R) \cong Q/QJ(R)$, prove that $P \cong Q$.
20. Let R be right perfect. Prove that a right R-module A is projective if and only if the map $A \otimes_R J(R) \to A$ is injective.

21. If R is right perfect, prove that every projective right R-module is isomorphic to a direct sum of copies of right ideals of R.

22. If R is right perfect, prove that r.gl.dim.$(R) \leqq$ l.gl.dim.(R).

B. Direct Products of Flat and Projective Modules

The main concern of this section is to characterize those rings R for which all direct products of projective right R-modules are projective. We proceed by first solving a related but easier problem: characterizing those rings R for which all direct products of flat right R-modules are flat. For this purpose, we begin by considering some relationships between the tensor product and the direct product.

Consider any *left* R-module A and any collection $\{B_\gamma\}$ of *right* R-modules. For each λ, the projection map $\Pi B_\gamma \to B_\lambda$ induces a map $(\Pi B_\gamma)\otimes_R A \to B_\lambda \otimes_R A$, and these maps together induce a map $(\Pi B_\gamma) \otimes_R A \to \Pi(B_\gamma \otimes_R A)$, which we refer to as the "natural map" from $(\Pi B_\gamma) \otimes_R A$ into $\Pi(B_\gamma \otimes_R A)$. If there are only finitely many nonzero modules in the collection $\{B_\gamma\}$, then this natural map is of course an isomorphism, but in general it need not be either injective or surjective (Exercise 1).

A special case of this situation is when all the modules B_γ are copies of some fixed module B. Here the direct product ΠB_γ is identified with the set B^X of all functions from the index set X into B, and similarly $\Pi(B_\gamma \otimes_R A)$ is identified with $(B \otimes_R A)^X$. Thus the natural map is written $(B^X) \otimes_R A \to (B \otimes_R A)^X$. In case $B = R_R$, we further identify $B \otimes_R A$ with A, giving us a natural map $(R^X) \otimes_R A \to A^X$. [In this case, the natural map ϕ is defined by the rule $(\phi(t \otimes a))_\gamma = t_\gamma a$.]

Proposition 5.10 For any *left* R-module A, the following conditions are equivalent:

(a) A is finitely generated.

(b) The natural map $(\Pi B_\gamma) \otimes_R A \to \Pi(B_\gamma \otimes_R A)$ is surjective for all collections $\{B_\gamma\}$ of right R-modules.

(c) The natural map $(R^X) \otimes_R A \to A^X$ is surjective for all sets X.

Proof: (a) $\Rightarrow$ (b): Choose generators $a_1, \ldots, a_n$ for A. Given any element $x \in \Pi(B_\gamma \otimes_R A)$, we can write $x_\gamma = b_{1\gamma} \otimes a_1 + \cdots + b_{n\gamma} \otimes a_n$ for each γ, for suitable elements $b_{1\gamma}, \ldots, b_{n\gamma} \in B_\gamma$. For each i, the $b_{i\gamma}$ are the components of an element $b_i \in \Pi B_\gamma$, and we see that the element $b_1 \otimes a_1 + \cdots + b_n \otimes a_n$ in $(\Pi B_\gamma) \otimes_R A$ maps onto x via the natural map.

(b) $\Rightarrow$ (c) is clear.

(c) $\Rightarrow$ (a): Let X be the set A, and define $t \in A^X$ by setting $t_a = a$ for all $a \in X$. According to (c), there must exist an element $w_1 \otimes a_1 + \cdots +$

$w_n \otimes a_n$ in $(R^X) \otimes_R A$ which maps onto t under the natural map. Observing that $w_{1a}a_1 + \cdots + w_{na}a_n = a$ for all $a \in A$, we conclude that $a_1, \ldots, a_n$ generate A.□

Proposition 5.11 For any *left* R-module A, the following conditions are equivalent:

(a) A is finitely presented.

(b) The natural map $(\Pi B_\gamma)\otimes_R A \to \Pi(B_\gamma \otimes_R A)$ is bijective for all collections $\{B_\gamma\}$ of right R-modules.

(c) The natural map $(R^X)\otimes_R A \to A^X$ is bijective for all sets X.

Proof: (a) ⇒ (b): There exists a short exact sequence $0 \to K \to F \to A \to 0$ of left R-modules with F finitely generated free and K finitely generated. We construct a commutative diagram with exact rows as follows:

$$\begin{array}{ccccccc}
(\Pi B_\gamma) \otimes_R K & \longrightarrow & (\Pi B_\gamma) \otimes_R F & \longrightarrow & (\Pi B_\gamma) \otimes_R A & \longrightarrow & 0 \\
\downarrow f & & \downarrow g & & \downarrow h & & \\
\Pi(B_\gamma \otimes_R K) & \longrightarrow & \Pi(B_\gamma \otimes_R F) & \longrightarrow & \Pi(B_\gamma \otimes_R A) & \longrightarrow & 0
\end{array}$$

Inasmuch as F is a finite direct sum of copies of ${}_RR$, we infer that g is bijective. Also, f is surjective by 5.10, from which an easy diagram chase shows that h is bijective.

(b) ⇒ (c) is clear.

(c) ⇒ (a): In view of 5.10, A is finitely generated, hence there exists a short exact sequence $0 \to K \to F \to A \to 0$ of left R-modules with F finitely generated and free. Given any set X, we construct a commutative diagram with exact rows as follows:

$$\begin{array}{ccccccccc}
 & & (R^X) \otimes_R K & \to & (R^X) \otimes_R F & \to & (R^X) \otimes_R A & \longrightarrow & 0 \\
 & & \downarrow f & & \downarrow g & & \downarrow h & & \\
0 & \longrightarrow & K^X & \longrightarrow & F^X & \longrightarrow & A^X & \longrightarrow & 0
\end{array}$$

As above, g is bijective. Since h is bijective by (c), an easy diagram chase shows that f is surjective. Consequently, 5.10 says that K is finitely generated, whence A is finitely presented.□

In the same vein as 5.10 and 5.11 is the question of conditions under which the natural maps $(R^X)\otimes_R A \to A^X$ are all injective. One answer to this question is given in Exercise 2. (See also Exercise 3.)

Definition Recall that R is *left coherent* provided every finitely generated left ideal of R is finitely presented. As we have seen, left noetherian rings, left semihereditary rings, and regular rings are all left coherent. (See also Exercise 4.)

Theorem 5.12 The following conditions are equivalent:

(a) All direct products of flat *right* R-modules are flat.
(b) $(R^X)_R$ is flat for all sets X.
(c) R is *left* coherent.

Proof: (a) $\Rightarrow$ (b) is clear.

(b) $\Rightarrow$ (c): Let I be any finitely generated left ideal of R. Given any set X, it follows from (b) that the map $(R^X)\otimes_R I \to R^X$ is injective, from which we see that the natural map $(R^X)\otimes_R I \to I^X$ is injective. In light of 5.10, this map is surjective as well; hence 5.11 shows that I is finitely presented.

(c) $\Rightarrow$ (a): Let $\{B_\gamma\}$ be any collection of flat right R-modules. Any finitely generated left ideal I of R is finitely presented by (c), whence 5.11 says that the natural map $(\Pi B_\gamma)\otimes_R I \to \Pi(B_\gamma \otimes_R I)$ is bijective. Since each B_γ is flat, the maps $B_\gamma \otimes_R I \to B_\gamma$ are all injective; hence the map $(\Pi B_\gamma)\otimes_R I \to \Pi B_\gamma$ must be injective. It follows that the map $(\Pi B_\gamma)\otimes_R K \to \Pi B_\gamma$ is injective for all left ideals K, and thus ΠB_γ is flat.□

Definition An R-module A is *torsionless* if A can be embedded in a direct product of copies of R. [Equivalently, A is torsionless if and only if the natural map of A into its double dual $\mathrm{Hom}_R(\mathrm{Hom}_R(A,R),R)$ is injective.] For example, all projective modules are torsionless, and all direct products of copies of ideals of R are torsionless. (See also Exercise 5.)

Theorem 5.13 All torsionless *right* R-modules are flat if and only if R is *left* semihereditary.

Proof: If all torsionless right R-modules are flat, then $(R^X)_R$ is flat for all sets X, whence 5.12 shows that R must be left coherent. Also, all right ideals of R are torsionless and thus flat, from which we infer that all left ideals of R are flat as well. Now all finitely generated left ideals of R are flat and finitely presented, hence projective, by 5.3. Therefore R is left semihereditary.

Conversely, assume that R is left semihereditary, and let A be any torsionless right R-module. We may assume that $A \leqq (R^X)_R$ for some set X. Observing that R is left coherent, we see from 5.12 that $(R^X)_R$ is flat. Thus A is a submodule of a flat module, from which we conclude that A must be flat.□

We now turn to the problem of characterizing those rings R for which all direct products of projective right R-modules are projective. The main difficulty is to show that such rings are right perfect, which is the purpose of the following theorem.

Theorem 5.14 Let X be an infinite set, and let $R_\gamma = R_R$ for all $\gamma \in X$. If $\prod_{\gamma \in X} R_\gamma$ is isomorphic to a direct summand of a right R-module $C = \bigoplus_{\lambda \in W} C_\lambda$ such that $\operatorname{card}(C_\lambda) \leqq \operatorname{card}(X)$ for all $\lambda \in W$, then R is a right perfect ring.

Proof: If not, then R has an infinite strictly decreasing sequence $Ra_0 > Ra_1 > \cdots$ of principal left ideals. We may clearly replace a_0 by 1, so that the sequence begins with $Ra_0 = R$.

Inasmuch as $\theta = \operatorname{card}(X)$ is infinite, X must have subsets $X_0 = X \supset X_1 \supset X_2 \supset \cdots$ such that $\operatorname{card}(X_n) = \theta$ for each n and $\cap X_n$ is empty. Set $A_n = \prod_{\gamma \in X_n} R_\gamma$ for each $n = 0,1, \ldots,$ and note that $A_n \cong A_0$ for each n. Also, we identify each A_n with the direct summand

$$\{x \in A_0 \mid x_\gamma = 0 \text{ for } \gamma \in X - X_n\} \subseteq A_0$$

We are given that A_0 is isomorphic to a direct summand of C; hence there exist homomorphisms $f : A_0 \to C$ and $g : C \to A_0$ such that gf is the identity map on A_0. For each $\lambda \in W$, let $p_\lambda : A_0 \to C_\lambda$ be the composition of f with the projection $C \to C_\lambda$.

Claim I: For any $x \in A_0$, $p_\lambda x = 0$ for all but finitely many $\lambda \in W$.
This is clear since C is the direct *sum* of the C_λ.

Claim II: If $x \in A_0$ and $a \in R$ such that $p_\lambda x \in C_\lambda a$ for all $\lambda \in W$, then $x \in A_0 a$.

For each $\lambda \in W$, choose an element $y_\lambda \in C_\lambda$ such that $y_\lambda a = p_\lambda x$, making sure that $y_\lambda = 0$ whenever $p_\lambda x = 0$. In view of Claim I, $y_\lambda = 0$ for all but finitely many $\lambda \in W$, hence the y_λ are the components of an element $y \in C$. Clearly $ya = fx$, whence $x = gfx = (gy)a$.

Given $n,k \geqq 0$, $\gamma \in X$, and $\lambda \in W$, we define the following abelian groups: $A_{nk} = (A_n a_k)/(A_n a_{k+1})$, $R_{\gamma k} = (R_\gamma a_k)/(R_\gamma a_{k+1})$, and $C_{\lambda k} = (C_\lambda a_k)/(C_\lambda a_{k+1})$. For elements $x \in A_n a_k$, we write $\bar{x}$ for the coset $x + A_n a_{k+1}$ in A_{nk}. We observe that p_λ induces a group homomorphism $p_{\lambda k} : A_{0k} \to C_{\lambda k}$. By identifying A_{nk} with the obvious submodule of A_{0k}, we may apply $p_{\lambda k}$ to A_{nk} also.

Claim III: $\operatorname{card}(A_{nk}) > \theta$ for all $n,k \geqq 0$.

For each $\gamma \in X_n$, $R_{\gamma k} \cong (Ra_k)/(Ra_{k+1})$, which has at least two elements. Inasmuch as $A_{nk} \cong \prod_{\gamma \in X_n} R_{\gamma k}$ and $\operatorname{card}(X_n) = \theta$, we see that $\operatorname{card}(A_{nk}) \geqq 2^\theta > \theta$.

Claim IV: If $n,k \geqq 0$ and $y \in A_{nk}$ is nonzero, then $p_{\delta k} y \neq 0$ for some $\delta \in W$.

Write $y = \bar{x}$ for some $x \in A_n a_k$. If $p_{\lambda k} y = 0$ for all $\lambda \in W$, then $p_\lambda x \in C_\lambda a_{k+1}$ for all $\lambda \in W$; hence by Claim II we obtain $x \in A_0 a_{k+1}$. As a result, $y = \bar{x} = 0$, which is false.

Claim V: If $n,k \geqq 0$ and W' is a finite subset of W, then there exist elements $x \in A_n a_k$ and $\delta \in W - W'$ such that $p_\delta x \notin C_\delta a_{k+1}$.

The maps $p_{\lambda k}$ (for $\lambda \in W'$) induce a map $h : A_{nk} \to \bigoplus_{\lambda \in W'} C_{\lambda k}$. Inasmuch as $\text{card}(C_{\lambda k}) \leqq \text{card}(C_\lambda) \leqq \theta$ for all $\lambda \in W'$, we infer that $\text{card}(\bigoplus_{\lambda \in W'} C_{\lambda k}) \leqq \theta$ (since W' is finite and θ is infinite). On the other hand, $\text{card}(A_{nk}) > \theta$ by Claim III, whence h cannot be injective. Thus there exists a nonzero element $y \in A_{nk}$ such that $hy = 0$, that is, $p_{\lambda k} y = 0$ for all $\lambda \in W'$. According to Claim IV, $p_{\delta k} y \neq 0$ for some $\delta \in W$, and clearly $\delta \notin W'$. Choosing an element $x \in A_n a_k$ for which $\bar{x} = y$, we conclude that $p_\delta x \notin C_\delta a_{k+1}$.

We now inductively define sequences $x_0, x_1, \ldots \in A_0$ and $\lambda(0), \lambda(1), \ldots \in W$ such that (a) $x_n \in A_n a_n$ for all n; (b) $p_{\lambda(n)} x_n \notin C_{\lambda(n)} a_{n+1}$ for all n; and (c) $p_{\lambda(n)} x_i = 0$ whenever $i < n$.

Using Claim V with $n = k = 0$ and W' empty, we first obtain elements $x_0 \in A_0 a_0$ and $\lambda(0) \in W$ such that $p_{\lambda(0)} x_0 \notin C_{\lambda(0)} a_1$. Now let $n > 0$, and assume we have elements $x_0, x_1, \ldots, x_{n-1} \in A_0$ and $\lambda(0), \ldots, \lambda(n-1) \in W$ satisfying (a) to (c). According to Claim I, the set

$$W' = \{\lambda \in W \mid p_\lambda x_i \neq 0 \text{ for some } i = 0,1, \ldots, n-1\}$$

must be finite; hence Claim V says that there exist elements $x_n \in A_n a_n$ and $\lambda(n) \in W - W'$ such that $p_{\lambda(n)} x_n \notin C_{\lambda(n)} a_{n+1}$. Since $\lambda(n) \notin W'$, we also have $p_{\lambda(n)} x_i = 0$ for $i = 0, \ldots, n-1$; hence the induction works. In light of (b) and (c), we note that $\lambda(i) \neq \lambda(n)$ whenever $i < n$.

Consider any $\gamma \in X$. Inasmuch as $\cap X_n$ is empty, $\gamma \notin X_n$ for some n. Now $x_j \in A_j a_j \subseteq A_n$ for all $j \geqq n$, whence $x_{j\gamma} = 0$ for all $j \geqq n$. Thus we can define an element $z_\gamma = \sum_{j=0}^{\infty} x_{j\gamma}$ in R_γ. Having defined $z_\gamma \in R_\gamma$ for all $\gamma \in X$, we obtain an element $z \in A_0$.

Next consider any $n \geqq 0$. Given $\gamma \in X$, we have $x_j \in A_j a_j \subseteq A_j a_{n+1}$ for all $j > n$ and thus $x_{j\gamma} \in R_\gamma a_{n+1}$ for all $j > n$, whence $z_\gamma = x_{0\gamma} + \cdots + x_{n\gamma} + y_\gamma a_{n+1}$ for some $y_\gamma \in R_\gamma$. This gives us an element $y \in A_0$ such that $z = x_0 + \cdots + x_n + y a_{n+1}$. Recalling that $p_{\lambda(n)} x_i = 0$ for $i < n$, we see that $p_{\lambda(n)} z = p_{\lambda(n)} x_n + (p_{\lambda(n)} y) a_{n+1}$. Inasmuch as $p_{\lambda(n)} x_n \notin C_{\lambda(n)} a_{n+1}$, we conclude that $p_{\lambda(n)} z \neq 0$.

Since $\lambda(0), \lambda(1), \ldots$ are all distinct, we thus have an element $z \in A_0$ with $p_\lambda z \neq 0$ for infinitely many λ, which contradicts Claim I.□

Theorem 5.15 The following conditions are equivalent:

(a) All direct products of projective *right* R-modules are projective.

(b) $(R^X)_R$ is projective for all sets X.

(c) R is *right* perfect and *left* coherent.

Proof: (a) $\Rightarrow$ (b) is clear.

(b) $\Rightarrow$ (c): In view of 5.12, R must be left coherent. Now choose an infinite set X with $\text{card}(X) \geqq \text{card}(R)$. Since $(R^X)_R$ is projective by (b), it must be isomorphic to a direct summand of a module $C = \oplus C_\lambda$, where each $C_\lambda = R_R$. According to 5.14, R is right perfect.

(c) $\Rightarrow$ (a): If A is any direct product of projective right R-modules, then A is also a direct product of flat right R-modules, and so A is flat, by 5.12. Since R is right perfect, A must be projective.□

For example, 5.9 shows that any left artinian ring is left noetherian and right perfect; hence such a ring satisfies 5.15(c). The converse implication is not true in general (Exercise 6), but it does hold for commutative rings (Exercise 7).

Exercises

1. If $B_n = \mathbf{Z} \oplus (\mathbf{Z}/n\mathbf{Z})$ for $n = 1, 2, \ldots$, show that the natural map $(\Pi B_n) \otimes_{\mathbf{Z}} \mathbf{Q} \rightarrow \Pi(B_n \otimes_{\mathbf{Z}} \mathbf{Q})$ is neither injective nor surjective.
2. Given ${}_RA$, prove that the natural map $(R^X) \otimes_R A \rightarrow A^X$ is injective for all sets X if and only if for every finitely generated $B \leqq A$, the inclusion map $B \rightarrow A$ factors through a finitely presented module.
3. Let R be regular. Given ${}_RA$, prove that the natural map $(R^X) \otimes_R A \rightarrow A^X$ is injective for all sets X if and only if all finitely generated submodules of A are projective.
4. If R is left noetherian and X is any set of indeterminates, prove that $R[X]$ is left coherent.
5. Prove that an essentially finitely generated $\mathbf{Z}$-module is torsionless if and only if it is free.
6. Let F be a field, V an infinite-dimensional vector space over F, $R = \begin{pmatrix} F & 0 \\ V & F \end{pmatrix}$. Show that all direct products of projective right or left R-modules are projective, but that R is neither right nor left artinian.
7. If R is commutative, prove that all direct products of projective R-modules are projective if and only if R is artinian.
8. Let R be commutative and regular. Prove that R is self-injective if and only if the natural map $(R^X) \otimes_R (R^X) \rightarrow R^{X \times X}$ is injective for all sets X. (*Hint*: 3.12.)
9. If R is the ring described in 3.11, show that the natural map $(R^X) \otimes_R (R^X) \rightarrow R^{X \times X}$ is not injective for any infinite set X.
10. If R is left coherent, prove that every finitely generated submodule of a finitely presented left R-module is finitely presented.
11. Prove that R is left coherent if and only if $\{r \in R \mid rx \in I\}$ is finitely generated for all $x \in R$ and all finitely generated left ideals I of R.
12. If R is left coherent, prove that the intersection of finitely many finitely generated left ideals of R is finitely generated.

13. If R is left coherent and I is a two-sided ideal of R such that ${}_RI$ is finitely generated, prove that R/I is left coherent.
14. With R,B as in Exercise 3.A.4, show that R is left coherent but that R/B is not.
15. If R and S are Morita-equivalent rings, prove that R is left coherent if and only if S is left coherent.
16. Prove that every torsionless **Z**-module has ACC on cyclic submodules.
17. Show that no infinite direct product of copies of **Z** can be free.
18. If every right R-module is a direct sum of countably generated right R-modules, show that R is right perfect.
19. Show that all torsionless right R-modules are projective if and only if R is left semihereditary and right perfect.
20. Let D be a division ring, V an infinite-dimensional vector space over D, $R = \text{End}_D(V)$. Show that R is right and left coherent. Show that ${}_R(R^X)$ is projective for all countable sets X, but that R is not left perfect.
21. Use 5.14 to prove that every right noetherian right self-injective ring is right artinian.

C. Projective Nonsingular Modules

Proposition 5.16 Let $Z_r(R) = 0$. Then all nonsingular *right* R-modules are flat if and only if $(S°R)_R$ is flat and R is *left* semihereditary.

Proof: If all nonsingular right R-modules are flat, then in particular $(S°R)_R$ must be flat. Observing that all torsionless right R-modules are nonsingular and thus flat, we see from 5.13 that R is left semihereditary.

Conversely, assume that $(S°R)_R$ is flat and that R is left semihereditary. Given any nonsingular right R-module A, we use the regularity of $S°R$ to see that $S°A$ is a flat right $S°R$-module. Inasmuch as $(S°R)_R$ is flat, we infer that $(S°A)_R$ is flat. Therefore A is a submodule of a flat module, from which we conclude that A must be flat.□

For example, every semihereditary commutative integral domain satisfies the conditions of 5.16 (Exercise 1). (See also Exercises 2 and 3.) We note that the conditions of 5.16 are not right-left symmetric (Exercise 4).

Theorem 5.17 If $Z_r(R) = 0$, then the following conditions are equivalent:

(a) Every finitely generated nonsingular right R-module can be embedded in a free right R-module.

(b) $(S°R)_R$ is flat, and the multiplication map $S°R \otimes_R S°R \to S°R$ is an isomorphism.

Proof: Set $Q = S°R$.

(a) ⇒ (b): According to 3.9, it suffices to show that if $x \in Q$ and $J = \{r \in R \mid rx \in R\}$, then $QJ = Q$.

In view of (a), there exists a monomorphism f from $A = R + xR$ into some $F = R_1 \oplus \cdots \oplus R_k$, where each $R_i = R_R$. For $i = 1, \ldots, k$, let p_i denote the projection $F \to R_i$, and note that the restriction of $p_i f$ to R is just left multiplication by the element $r_i = p_i f(1) \in R$. Thinking of $p_i f$ and left multiplication by r_i as maps from A into Q, we see by 2.1 that $p_i f$ must be left multiplication by r_i. In particular, $r_i x = p_i f(x) \in R$, whence $r_i \in J$.

Inasmuch as Q_R is injective, the isomorphism $f^{-1}: fA \to A$ extends to a map $g: F \to Q$. Then gf is just the inclusion map $A \to Q$, and consequently $gf(1) = 1$.

For each $i = 1, \ldots, k$, let j_i denote the injection map $R_i \to F$, and note that $gj_i : R_R \to Q_R$ is just left multiplication by the element $q_i = gj_i(1) \in Q$. Now $j_1 p_1 + \cdots + j_k p_k = 1_F$, hence we see that $gf = (gj_1)(p_1 f) + \cdots + (gj_k)(p_k f)$, which is just left multiplication by the element $q_1 r_1 + \cdots + q_k r_k$. Inasmuch as $gf(1) = 1$, we obtain $q_1 r_1 + \cdots + q_k r_k = 1$, and consequently $QJ = Q$.

(b) $\Rightarrow$ (a): If A is any finitely generated nonsingular right R-module, then in view of 2.12 we may assume that $A \leqq Q_1 \oplus \cdots \oplus Q_k$, where each $Q_i = Q_R$. We may choose finitely generated R-submodules $A_i \leqq Q_i$ for each i such that $A \leqq A_1 \oplus \cdots \oplus A_k$. Since it suffices to embed each A_i in a free right R-module, we may thus assume, without loss of generality, that A is a finitely generated submodule of Q_R.

Choose generators $x_1, \ldots, x_n$ for A, and set $J_i = \{r \in R \mid rx_i \in R\}$ for each $i = 1, \ldots, n$. According to 3.9, $QJ_i = Q$ for each i, whence

$$Q \otimes_R [(R/J_1) \oplus \cdots \oplus (R/J_n)] = 0$$

If $J = J_1 \cap \cdots \cap J_n$, then there is a monomorphism of R/J into $(R/J_1) \oplus \cdots \oplus (R/J_n)$. Inasmuch as Q_R is flat, this induces a monomorphism of $Q \otimes_R (R/J)$ into $Q \otimes_R [(R/J_1) \oplus \cdots \oplus (R/J_n)]$, from which we see that $Q \otimes_R (R/J) = 0$. Thus $QJ = Q$, hence there exist elements $r_1, \ldots, r_t \in J$ and $q_1, \ldots, q_t \in Q$ such that $q_1 r_1 + \cdots + q_t r_t = 1$.

For each $j = 1, \ldots, t$, we have $r_j x_1, \ldots, r_j x_n \in R$ and consequently $r_j A \leqq R$. Thus left multiplication by r_j defines a homomorphism of A into R_R. Together, these maps induce a homomorphism $f: A \to F$, where F is a direct sum of t copies of R_R. Inasmuch as $q_1 r_1 + \cdots + q_t r_t = 1$, we see that $\ker f = 0$, hence f is an embedding of A into the free module F. □

Note the similarity of 5.17(b) and 3.10(a). In case R is commutative, these conditions coincide, as in Exercise 5. For example, these conditions are satisfied by all commutative integral domains. (See also Exercise 3.D.14 and Exercise 6.) Exercises 7 and 8 show that the conditions of 5.17 are unrelated to those of 5.16. We also note that the conditions of 5.17 are not right-left symmetric (Exercise 9).

Theorem 5.18 If $Z_r(R) = 0$, then the following conditions are equivalent:

(a) All finitely generated nonsingular right R-modules are projective.

(b) R is right semihereditary, $(S^\circ R)_R$ is flat, and the multiplication map $S^\circ R \otimes_R S^\circ R \to S^\circ R$ is an isomorphism.

(c) R is right and left semihereditary, ${}_RR \leqq_e {}_R(S^\circ R)$, and the multiplication map $S^\circ R \otimes_R S^\circ R \to S^\circ R$ is an isomorphism.

Proof: Set $Q = S^\circ R$.

(a) $\Rightarrow$ (b): Since all finitely generated right ideals of R are nonsingular and thus projective, R is right semihereditary. Also, 5.17 shows that Q_R is flat and that the multiplication map $Q \otimes_R Q \to Q$ is an isomorphism.

(b) $\Rightarrow$ (a): If A is any finitely generated nonsingular right R-module, then by 5.17, A is isomorphic to a submodule of a free right R-module. Inasmuch as R is right semihereditary, A must be projective.

(a) $\Rightarrow$ (c): According to (b), R is right semihereditary and the multiplication map $Q \otimes_R Q \to Q$ is an isomorphism. If A is any nonsingular right R-module, then all finitely generated submodules of A are projective and thus flat, from which it follows that A is flat. Now 5.16 says that R is left semihereditary.

Consider any nonzero $x \in Q$, and set $J = \{r \in R \mid rx \in R\}$. In view of (b), we see from 3.9 that $QJ = Q$; hence $Jx \neq 0$ and so $Rx \cap R \neq 0$. Thus ${}_RR \leqq_e {}_RQ$.

(c) $\Rightarrow$ (b): According to 1.27, $Z_l(R) = 0$. Inasmuch as ${}_RR \leqq_e {}_RQ$, we thus obtain $Z({}_RQ) = 0$.

We only need to prove that Q_R is flat, and for this it suffices to show that every finitely generated $A \leqq Q_R$ is flat. According to 5.13, all torsionless right R-modules are flat; hence we need only show that A is torsionless.

Choose generators $x_1, \ldots, x_n$ for A, and set $J = \{r \in R \mid rx_i \in R \text{ for all } i\}$. Since ${}_RR \leqq_e {}_RQ$, we must have $J \leqq_e {}_RR$. Inasmuch as $Z({}_RQ) = 0$, it follows that the right annihilator of J in Q is zero. Thus for any nonzero $x \in A$ there exists an element $r \in J$ such that $rx \neq 0$; hence left multiplication by r defines a homomorphism $f: A \to R_R$ such that $fx \neq 0$. Therefore A is torsionless. □

For example, all semihereditary commutative integral domains satisfy the conditions of 5.18 (Exercise 10). Another class of examples is provided by 3.12: If R is any regular, right self-injective ring, then all finitely generated nonsingular right R-modules are projective. As noted in the proof above, the conditions of 5.18 imply those of 5.16 as well as those of 5.17. Neither converse holds in general (Exercises 11 and 12), and in fact the conditions of 5.16 and 5.17 together do not imply those of 5.18 (Exercise 13). (See also Exercise 14.) Exercise 15 shows that the conditions of 5.18 are not right-left symmetric. (See also Exercise 16.)

Proposition 5.19 If $Z_r(R) = 0$, then any essentially finitely generated projective right R-module P is finitely generated.

Proof: There exist elements $x_1, \ldots, x_n \in P$ such that $x_1R + \cdots + x_nR \leqq_e P$.

According to the Dual Basis Lemma, there exist elements $\{y_\gamma \mid \gamma \in X\}$ in P and maps $\{f_\gamma \mid \gamma \in X\}$ in $\operatorname{Hom}_R(P, R_R)$ such that for all $x \in P$, $f_\gamma x = 0$ for all but finitely many $\gamma \in X$ and $x = \sum y_\gamma(f_\gamma x)$. Now the set

$$Y = \{\gamma \in X \mid f_\gamma x_i \neq 0 \text{ for some } i = 1, \ldots, n\}$$

must be finite. For $\gamma \in X - Y$, we thus have $f_\gamma(x_1R + \cdots + x_nR) = 0$, whence 2.1 shows that $f_\gamma = 0$. Consequently, $x = \sum_{\gamma \in Y} y_\gamma(f_\gamma x)$ for all $x \in P$, whence P is generated by the finite set $\{y_\gamma \mid \gamma \in Y\}$. □

Corollary 5.20 If R is right hereditary and R_R is finite-dimensional, then R is right noetherian.

Proof: By 1.27, $Z_r(R) = 0$. Any right ideal I of R is projective because R is right hereditary, and I is essentially finitely generated by 3.13, whence 5.19 shows that I is finitely generated. Therefore R is right noetherian. □

Theorem 5.21 If $Z_r(R) = 0$, then the following conditions are equivalent:

(a) All nonsingular right R-modules are projective.

(b) R is right perfect and *left* semihereditary, and $(S^\circ R)_R$ is flat.

(c) R is right hereditary and right artinian, and $(S^\circ R)_R$ is flat.

(d) R is right hereditary, R_R is finite-dimensional, and R has a two-sided ideal K such that K_R is injective and ${}_RK \leqq_e {}_RR$.

Proof: Set $Q = S^\circ R$.

(a) ⇒ (b): In view of 5.16, R is left semihereditary and Q_R is flat. Inasmuch as any direct product of projective right R-modules is nonsingular and thus projective, 5.15 shows that R is right perfect.

(b) ⇒ (a): According to 5.16, all nonsingular right R-modules are flat. Since R is right perfect, we also have that all flat right R-modules are projective.

(a) ⇒ (c): Since all right ideals of R are nonsingular and thus projective, R is right hereditary. We also see from (b) that Q_R is flat.

We next use 5.15 to show that Q is right perfect. Thus let X be any set, and put $A = (Q^X)_Q$. As a right R-module, A is nonsingular and thus projective. Choosing a Q-epimorphism $f : F \to A$ with F_Q free, it follows that there is at least an R-homomorphism $g : A \to F$ such that $fg = 1_A$. Since F_R is nonsingular, 2.7 says that g is also a Q-homomorphism, whence A_Q is projective. Now 5.15 shows that Q is right perfect; hence $Q/J(Q)$ is a semisimple ring. However, $J(Q) = 0$ because Q is regular, and consequently Q must be semisimple.

According to 3.17, R_R is finite-dimensional. Inasmuch as R is also right hereditary, we find that R is right noetherian by 5.20. Finally, since R is right perfect by (b), we conclude from 5.9 that R is right artinian.

(c) $\Rightarrow$ (d): We are given that R is right hereditary, and since R_R is artinian it must also be finite-dimensional. As a result, 3.17 says that Q is a semisimple ring. Also, R is right perfect by 5.9 and Q_R is flat by (c), whence Q_R must be projective.

Let K denote the two-sided ideal $\{r \in R \mid rQ \leqq R\}$, and note that $KQ = K$. Inasmuch as Q is semisimple, K_Q is a direct summand of Q_Q, from which it follows that K_R is injective.

Given any nonzero $a \in R$, we infer from the projectivity of Q_R that there is a homomorphism $f : Q_R \to R_R$ such that $fa \neq 0$. Considering f as a map of $Q_R \to Q_R$, it follows from 2.7 that f is also a right Q-homomorphism, whence f must be left multiplication by some $r \in Q$. Since $rQ = fQ \leqq R$, we have $r \in K$; hence ra is a nonzero element of $Ra \cap K$. Therefore ${}_RK \leqq_e {}_RR$.

(d) $\Rightarrow$ (a): Using 3.17 once again, we see that Q is a semisimple ring; hence Q_Q is a direct sum of minimal right ideals. We claim that Q_R is projective, which will follow from showing that every minimal right ideal A of Q is projective as a right R-module.

In view of 2.9, $S^\circ A = A$ and every member of $L^*(A)$ is a Q-submodule of A. Since A_Q is simple, we thus obtain $L^*(A) = \{0,A\}$. Inasmuch as $R_R \leqq_e Q_R$, there exists a nonzero element $a \in A \cap R$. We have ${}_RK \leqq_e {}_RR$ as well, hence there exists an element $r \in R$ such that ra is a nonzero element of K. Then left multiplication by r defines a nonzero map of aR into the injective module K_R, and this map must extend to a nonzero map $f : A_R \to K_R$. Now $\ker f \in L^*(A)$ and $\ker f \neq A$, whence $\ker f = 0$. Thus A_R is isomorphic to a right ideal of R, which is projective because R is right hereditary.

Therefore Q_R is projective, as claimed. Given any nonsingular right R-module B, we have $(S^\circ B)_Q$ projective because Q is semisimple, from which we infer that $(S^\circ B)_R$ is projective. Inasmuch as R is right hereditary, B must thus be projective. □

The conditions of 5.21 obviously imply those of 5.18, and it is easy to see that the converse fails. For example, let V be an infinite-dimensional vector space over a division ring D, and set $R = \mathrm{End}_D(V)$. Since R is regular and right self-injective by 2.23, 3.12 shows that all finitely generated nonsingular right R-modules are projective. However, R is not right artinian, hence 5.21 shows that not all nonsingular right R-modules are projective.

Certainly if R is semisimple then all nonsingular right R-modules are projective, and if R is commutative this is the only case which occurs (Exercise

17). In the noncommutative case, however, there exist non-semisimple rings R for which all nonsingular right R-modules are projective, as shown by the following proposition.

Proposition 5.22 Let T be a semisimple ring and n a positive integer. If R is the ring of all lower triangular $n \times n$ matrices over T, then R is a right and left nonsingular ring, and all nonsingular right and left R-modules are projective.

Proof: According to 4.9, R is right and left hereditary, and then 1.27 shows that R is right and left nonsingular.

The ring Q of all $n \times n$ matrices over T is obviously semisimple, whence Q is regular and right self-injective. Observing that $R_R \leqq_e Q_R$, we see from 2.11 that $S^\circ R = Q$. Consequently, 3.17 shows that R_R is finite-dimensional.

Let K denote the bottom row of R, and note that K is a two-sided ideal of R such that ${}_RK \leqq_e {}_RR$. Inasmuch as K is a right ideal of the semisimple ring Q, K_Q must be a direct summand of Q_Q, from which it follows that K_R is injective.

Now 5.21 says that all nonsingular right R-modules are projective, and by symmetry all nonsingular left R-modules are projective as well. □

As usual, we raise the question of whether the conditions of 5.21 are right-left symmetric. This time, however, the answer is yes, as shown by the following theorem.

Theorem 5.23 The following conditions are equivalent:

(a) $Z_r(R) = 0$ and all nonsingular right R-modules are projective.

(b) $Z_l(R) = 0$ and all nonsingular left R-modules are projective.

(c) R is right and left hereditary, right and left artinian, and the maximal right and left quotient rings of R coincide.

Proof: In light of 5.21, we see that in every case R is right and left semihereditary. According to 1.27, R is thus right and left nonsingular. We have the localization functor S° defined on right modules, and for the duration of this proof we use T° to denote the corresponding localization functor defined on left R-modules.

(a) ⇒ (b): Since R_R is finite-dimensional by 5.21, 3.17 says that $S^\circ R$ is a semisimple ring. In particular, $S^\circ R$ is regular and left self-injective. According to 5.18, ${}_RR \leqq_e {}_R(S^\circ R)$, whence 2.11 shows that $T^\circ R = S^\circ R$.

Inasmuch as R is right hereditary (by 5.21) and thus right coherent, we see from 3.7 that ${}_R(T^\circ R) = {}_R(S^\circ R)$ is flat. Also, R is right artinian by 5.21; hence 5.9 shows that R is left perfect. Since R is right semihereditary as well, it now follows from 5.21 that all nonsingular left R-modules are projective.

(b) ⇒ (a) by symmetry.

(a) $\Rightarrow$ (c): According to 5.21, R is right hereditary and right artinian, and by symmetry R is left hereditary and left artinian as well. As above, $S^\circ R = T^\circ R$, which by 2.31 shows that the maximal right and left quotient rings of R coincide.

(c) $\Rightarrow$ (a): Using 2.31 again, we obtain $S^\circ R = T^\circ R$. Since R is left hereditary and thus left coherent, it now follows from 3.7 that $(S^\circ R)_R = (T^\circ R)_R$ is flat. Inasmuch as R is also right hereditary and right artinian, we conclude from 5.21 that all nonsingular right R-modules are projective. □

The remainder of this chapter is devoted to representing rings for which all nonsingular modules are projective in terms of lower triangular matrix rings over division rings. Several preparatory results are needed first.

Definition Recall that any module A which is both artinian and noetherian has at least one composition series $A_0 = 0 < A_1 < \cdots < A_n = A$. The Jordan-Hölder Theorem implies that the integer n is the same for all composition series of A. It is known as the *length of* A, and we denote it $l(A)$.

Definition A *uniserial* module is an artinian and noetherian module A which has exactly one composition series $A_0 = 0 < A_1 < \cdots < A_n = A$. Since any submodule of A can be fitted into a composition series, it follows that all submodules of A belong to the chain $A_0 < A_1 < \cdots < A_n$. As a consequence, all nonzero submodules of A are essential in A, so that A is either zero or uniform. Note also that if $A \neq 0$, then A has exactly one simple submodule, namely, A_1, and that $A_1 \leqq_e A$.

For example, 0 and all simple modules are clearly uniserial. If p is any prime number and n is any positive integer, then $\mathbf{Z}/p^n\mathbf{Z}$ is a uniserial $\mathbf{Z}$-module of length n. (See also Exercise 18.)

Lemma 5.24 Let $Z_r(R) = 0$, and assume that all nonsingular right R-modules are projective. Then any indecomposable nonsingular right R-module is uniserial.

Proof: Let A be any nonzero indecomposable nonsingular right R-module. For any $B \in L^*(A)$, A/B is nonsingular and thus projective, whence B is a direct summand of A. Since A is indecomposable, we obtain either $B = 0$ or $B = A$. Therefore $L^*(A) = \{0,A\}$, and consequently 3.24 shows that A is uniform.

Since A is a nonzero projective module, there is a nonzero homomorphism $f : A \to R_R$. Then $\ker f$ is a proper $\mathscr{S}$-closed submodule of A and so $\ker f = 0$, whence $A \cong fA \leqq R_R$. According to 5.21, R is right artinian, and by 5.9, R is right noetherian as well. As a result, A must have a composition series; hence we may proceed by induction on the length of A.

If $l(A) \leqq 1$, then A is either zero or simple, in which case A is clearly uniserial. Now let $l(A) = n > 1$, and assume that all indecomposable nonsingular right R-modules of length $n - 1$ are uniserial.

Since A has a composition series it has a maximal submodule M, and we claim that M is unique. If, on the contrary, A has another maximal submodule M', then we obtain a nonzero map

$$A \to A/(M \cap M') = [M/(M \cap M')] \oplus [M'/(M \cap M')] \to M/(M \cap M')$$

Inasmuch as A is nonsingular and thus projective, this map lifts to a nonzero map $g : A \to M$. Now ker g is a proper $\mathscr{S}$-closed submodule of A; hence ker $g = 0$ and A is isomorphic to a submodule of M. Inasmuch as $l(A) = n$ while $l(M) = n - 1$, this is impossible.

Therefore M is unique, as claimed; hence any composition series for A must include M. Since A is uniform, M is an indecomposable nonsingular right R-module of length $n - 1$; hence the induction hypothesis says that M is uniserial. We now conclude that A must be uniserial. □

Proposition 5.25 Let A be a nonzero injective uniserial right R-module with composition series $A_0 = 0 < A_1 < \cdots < A_n = A$, and set $P = A_1 \oplus \cdots \oplus A_n$. If P is projective, then $D = \mathrm{End}_R(A)$ is a division ring and $W = \mathrm{End}_R(P)$ is isomorphic to the ring of all lower triangular $n \times n$ matrices over D.

Proof: Since the only submodules of A are $A_0, \ldots, A_n$, we see from the projectivity of P that all submodules of A are projective. Given any nonzero $f \in D$, it follows that fA is projective, whence ker f is a direct summand of A. Inasmuch as A is uniform, we obtain ker $f = 0$; hence $fA \cong A$. Then $l(fA) = l(A)$, from which we infer that $fA = A$, and so f is an isomorphism. Therefore D is a division ring.

Set $W_{ij} = \mathrm{Hom}_R(A_j, A_i)$ for all $i,j = 1, \ldots, n$. We first claim that $W_{ij} = 0$ whenever $i < j$.

Given any $f \in W_{ij}$, we use the projectivity of fA_j to see that ker f is a direct summand of A_j. Since A is uniform, either ker $f = 0$ or ker $f = A_j$. For $i < j$, we have $l(fA_j) \leqq l(A_i) < l(A_j)$, whence f cannot be injective. Thus ker $f = A_j$ and $f = 0$, which establishes the claim.

Secondly, we claim that whenever $j \leqq i$, $DA_j \leqq A_i$ and the restriction map $\phi_{ij} : D \to W_{ij}$ is a group isomorphism.

Given any $f \in D$, we have $l(fA_j) \leqq l(A_j) = j \leqq i$. Since the only submodules of A with length at most i are $A_0, \ldots, A_i$, it follows that $fA_j \leqq A_i$. Thus $DA_j \leqq A_i$, so that the restriction map ϕ_{ij} does indeed map D into W_{ij}. Inasmuch as A is injective, ϕ_{ij} must be surjective. If $f \in D$ with $\phi_{ij}(f) = 0$, then f kills the nonzero module A_j and so cannot be invertible in D, whence $f = 0$. Thus ϕ_{ij} is an isomorphism.

We may identify W with the ring of all $n \times n$ matrices with i,j-th entries from W_{ij}. If L denotes the ring of all lower triangular $n \times n$ matrices over D, then we can define a ring homomorphism $\phi : L \to W$ by setting $(\phi x)_{ij} = \phi_{ij}(x_{ij})$ whenever $j \leqq i$, and $(\phi x)_{ij} = 0$ whenever $j > i$. Inasmuch as ϕ_{ij} is an isomorphism for $j \leqq i$ and $W_{ij} = 0$ for $j > i$, we conclude that ϕ is an isomorphism.□

Lemma 5.26 Let $Z_r(R) = 0$, and assume that all cyclic nonsingular right R-modules are projective. If R is indecomposable (as a ring), then all minimal right ideals of R are isomorphic.

Proof: We may assume that R has at least one minimal right ideal A. Let $\mathscr{F} = \{F \leqq (S^\circ R)_R \mid F \cong A\}$, and set $H = \Sigma\mathscr{F}$. For any $F \in \mathscr{F}$, we have $F \cap R \neq 0$ because $R_R \leqq_e (S^\circ R)_R$; hence it follows from the simplicity of F that $F = F \cap R \leqq R$. Consequently, H is a right ideal of R. Given $F \in \mathscr{F}$ and $x \in S^\circ R$, we either have $xF = 0$ or else $xF \cong F \cong A$, so that $xF \leqq H$ in either case. As a result, we find that H is also a left ideal of $S^\circ R$.

Given any $x \in S^\circ R$, we thus have $xH \leqq H$. Since $S^\circ H/H$ is singular and $S^\circ H \in L^*(S^\circ R)$, it follows that $x(S^\circ H) \leqq S^\circ H$ as well. Thus $S^\circ H$ is a two-sided ideal of $S^\circ R$; hence $K = R \cap S^\circ H$ is a two-sided ideal of R. Note that $H_R \leqq_e K_R$.

Inasmuch as $S^\circ H \in L^*(S^\circ R)$, we see that $K \in L^*(R_R)$. Then R/K is a cyclic nonsingular right R-module and so is projective; hence $K = eR$ for some idempotent $e \in R$. Observing that $H_R \leqq_e K_R \leqq_e e(S^\circ R)$ and that $e(S^\circ R) \in L^*(S^\circ R)$, we see from 2.5 that $e(S^\circ R) = S^\circ H$. Thus $e(S^\circ R)$ is a two-sided ideal in the semiprime ring $S^\circ R$; hence 2.33 shows that e is central in $S^\circ R$. Now e is a central idempotent in the indecomposable ring R, from which it follows that either $e = 0$ or $e = 1$. Since $A \leqq H \leqq S^\circ H$, we must have $e \neq 0$, and consequently $K = eR = R$.

Therefore $H_R \leqq_e R_R$. Given any minimal right ideal C of R, we thus have $C \cap H \neq 0$, from which it follows that $C \leqq H$. Inasmuch as H is a direct sum of members of $\mathscr{F}$, there must exist a nonzero map $C \to F$ for some $F \in \mathscr{F}$. Since F and C are both simple, we see that $C \cong F \cong A$. Therefore all minimal right ideals of R are isomorphic to A.□

Definition *A* ring S is said to be a *full lower triangular matrix ring* over a ring T if there exists a positive integer n such that S is the ring of all lower triangular $n \times n$ matrices over T.

Theorem 5.27 Let $Z_r(R) = 0$, and assume that R is indecomposable (as a ring). Then all nonsingular right R-modules are projective if and only if R is Morita-equivalent to a full lower triangular matrix ring over a division ring.

Proof: Sufficiency is clear from 5.22. Now assume, conversely, that all nonsingular right R-modules are projective.

Since R_R is finite-dimensional by 5.21, 3.17 says that $S^\circ R$ is a semisimple ring; hence we may choose a minimal right ideal A of $S^\circ R$. Now A is a direct summand of $S^\circ R$, whence A_R is injective and $S^\circ A = A$. Consequently, A_R is uniform by 3.24 and so is indecomposable.

By 5.24, A is uniserial. Let $A_0 = 0 < A_1 < \cdots < A_n = A$ be the composition series for A, and set $P = A_1 \oplus \cdots \oplus A_n$. Now P is nonsingular and thus projective, and it is clear that P is finitely generated. According to 5.25, $\text{End}_R(P)$ is isomorphic to a full lower triangular matrix ring over a division ring; hence all that remains is to show that P is a generator in Mod-R.

Inasmuch as R_R is artinian (by 5.21), we infer that $R_R = J_1 \oplus \cdots \oplus J_k$ for some nonzero indecomposable right ideals J_i. For each i, J_i is uniserial by 5.24; hence it must contain a simple essential submodule S_i. Similarly, A_1 is a simple essential submodule of A, and we infer from 5.26 that $S_i \cong A_1$. Thus $E(S_i) \cong E(A_1) = A$, so that J_i is isomorphic to a submodule of A. Inasmuch as the only nonzero submodules of A are $A_1, \ldots, A_n$, we conclude that J_i must be isomorphic to a direct summand of P. Consequently, $R_R = J_1 \oplus \cdots \oplus J_k$ is isomorphic to a direct summand of P^k, and therefore P is a generator in Mod-R.□

Theorem 5.28 Let $Z_r(R) = 0$. Then all nonsingular right R-modules are projective if and only if R is Morita-equivalent to a finite direct product of full lower triangular matrix rings over division rings.

Proof: Sufficiency is clear from 5.22. Conversely, assume that all nonsingular right R-modules are projective. Since R is right artinian by 5.21, there is a ring decomposition $R = R_1 \times \cdots \times R_n$ with each R_i indecomposable. According to 5.27, each R_i is Morita-equivalent to a full lower triangular matrix ring T_i over a division ring D_i, and then R is Morita-equivalent to $T_1 \times \cdots \times T_n$.□

Exercises

1. If R is a semihereditary commutative integral domain, show that all nonsingular R-modules are flat.
2. Let R be commutative and nonsingular. Prove that all nonsingular R-modules are flat if and only if R is semihereditary. (*Hint*: Exercise 2.A.14.)
3. Let R be a semiprime right and left Goldie ring. If R is left semihereditary, prove that all nonsingular right R-modules are flat.
4. With R,Q as in 3.11, set $T = \begin{pmatrix} R & 0 \\ Q & Q \end{pmatrix}$. Show that $Z_r(T) = Z_l(T) = 0$. Show that all nonsingular right T-modules are flat, but that not all nonsingular left T-modules are flat. (*Hint*: Exercise 4.A.13.)
5. Let R be commutative and nonsingular. Show that every finitely generated nonsingular R-module can be embedded in a free R-module if and only if $\{r \in R \mid xr \in R\}$ is essentially finitely generated for all $x \in S^\circ R$.

6. If R is a semiprime right and left Goldie ring, prove that every finitely generated nonsingular right or left R-module can be embedded in a free R-module.
7. Let F be a field, $R = F[x,y]$. Show that $Z_r(R) = 0$ and that every finitely generated nonsingular R-module can be embedded in a free R-module, but that not all nonsingular R-modules are flat.
8. With R as in 3.11, show that all nonsingular R-modules are flat, but that not every finitely generated nonsingular R-module can be embedded in a free R-module.
9. Let F be a field, V an infinite-dimensional vector space over F, $R = \text{End}_F(V)$. Show that $Z_r(R) = Z_l(R) = 0$, and that every finitely generated nonsingular right R-module can be embedded in a free right R-module. Prove that not every finitely generated nonsingular left R-module can be embedded in a free left R-module.
10. If R is a semihereditary commutative integral domain, show that all finitely generated nonsingular R-modules are projective.
11. Find an example of a right nonsingular ring R such that all nonsingular right R-modules are flat, but not all finitely generated nonsingular right R-modules are projective.
12. Find an example of a right nonsingular ring R such that every finitely generated nonsingular right R-module can be embedded in a free right R-module, but not all finitely generated nonsingular right R-modules are projective.
13. Let $F \subset K$ be distinct fields, V any countably infinite-dimensional vector space over K, $T = \text{End}_K(V)$, M any maximal right ideal of T which contains $\text{soc}(T_T)$, $R = F + M$. Show that $Z_r(R) = 0$. Prove that all nonsingular right R-modules are flat, and that every finitely generated nonsingular right R-module can be embedded in a free right R-module. (*Hint*: Exercise 4.C.21.) Prove that not every finitely generated nonsingular right R-module is projective.
14. Let $Z_r(R) = 0$ and assume that R_R is finite-dimensional. If all nonsingular right R-modules are flat and every finitely generated nonsingular right R-module can be embedded in a free right R-module, prove that all finitely generated nonsingular right R-modules are projective. (*Hint*: Exercise 3.A.20.)
15. Let F be a field, V an infinite-dimensional vector space over F, $R = \text{End}_F(V)$. Show that $Z_r(R) = Z_l(R) = 0$. Show that all finitely generated nonsingular right R-modules are projective, but that not all finitely generated nonsingular left R-modules are projective.
16. Assume that R_R is finite-dimensional. If $Z_r(R) = 0$ and all finitely generated nonsingular right R-modules are projective, prove that $Z_l(R) = 0$ and that all finitely generated nonsingular left R-modules are projective.
17. Let R be commutative and nonsingular. If all nonsingular R-modules are projective, prove that R is semisimple.
18. Let R be a hereditary commutative integral domain which is not a field. If M is a maximal ideal of R and $n > 0$, prove that R/M^n is a uniserial R-module of length n.
19. Show that all uniserial modules are cyclic.

20. If $Z_r(R) = 0$ and all nonsingular right R-modules are projective, prove that the same holds for $\begin{pmatrix} R & 0 \\ S^\circ R & S^\circ R \end{pmatrix}$.

21. If $Z_r(R) = 0$ and all nonsingular right R-modules are projective, prove that every nonsingular right R-module is a direct sum of uniserial modules.

22. Let $Z_r(T) = 0$, M an essential semimaximal right ideal of T, R the idealizer of M in T. Prove that all finitely generated nonsingular right R-modules are projective if and only if all finitely generated nonsingular right T-modules are projective.

23. Let $Z_r(T) = 0$, M a generative essential right ideal of T, R a tame subidealizer of M in T. Prove that all nonsingular right R-modules are projective if and only if M is semimaximal in T, R is the idealizer of M in T, and all nonsingular right T-modules are projective.

24. Let $Z_r(R) = 0$. Prove that all nonsingular right R-modules are free if and only if R is a division ring.

6

Nonsingular Injective Modules

The purpose of this chapter is to study nonsingular injective modules over a nonsingular ring R, and to apply the information thus obtained to derive certain properties of the two-sided ideals in the ring $S^\circ R$. The first section is mainly concerned with prime ideals in $S^\circ R$. It is shown, for example, that the two-sided ideals in $S^\circ R/P$ are linearly ordered, for any prime ideal P. Section B develops some properties of modules which are not isomorphic to proper direct summands of themselves, for use in the succeeding section. Section C is mainly concerned with developing a multiplicity theory for nonsingular injective modules in the case when $S^\circ R$ is prime. A major consequence is that in this case the two-sided ideals of $S^\circ R$ are well-ordered.

A. Prime Ideals in $S^\circ R$

Perhaps the simplest question to ask about prime ideals in $S^\circ R$ is when 0 is a prime ideal, i.e., when $S^\circ R$ is a prime ring. This is answered by the following theorem.

Theorem 6.1 If $Z_r(R) = 0$ and $R \neq 0$, then the following conditions are equivalent:

(a) $S^\circ R$ is a prime ring.

(b) $S^\circ R$ is indecomposable (as a ring).

(c) If I,J are two-sided ideals of R such that $I \cap J = 0$ and $I \oplus J \in \mathscr{S}(R)$, then either $I = 0$ or $J = 0$.

(d) If I,J are nonzero right ideals of R, then there exist nonzero right ideals $I' \leqq I$ and $J' \leqq J$ such that $I' \cong J'$.

Proof: Set $Q = S^\circ R$, and note that $Q \neq 0$.

(a) $\Rightarrow$ (d): Since $S^\circ I$ and $S^\circ J$ are nonzero right ideals of Q, we must have $(S^\circ I)(S^\circ J) \neq 0$; hence $xy \neq 0$ for some $x \in S^\circ I$, $y \in S^\circ J$. Now xyQ is generated by an idempotent and so is projective. Also, left multiplication by x defines an epimorphism of yQ onto xyQ; hence we see that there must exist a monomorphism $f: xyQ \to yQ$. Inasmuch as $J \leqq_e S^\circ J$, we obtain $J \cap f(xyQ) \neq 0$; hence $f^{-1}J \neq 0$. We also have $I \leqq_e S^\circ I$, whence $I' = I \cap (f^{-1}J) \neq 0$. Since f is a monomorphism, it follows that $J' = fI'$ is a nonzero right ideal contained in J, and that $I' \cong J'$.

(d) $\Rightarrow$ (c): If I and J are both nonzero, then by (d) there exist nonzero right ideals $I' \leqq I$ and $J' \leqq J$ such that $I' \cong J'$. Now $I'J \leqq IJ \leqq I \cap J = 0$, and likewise $J'I = 0$. Since $I' \cong J'$, we obtain $I'I = 0$, and thus $I'(I \oplus J) = 0$. Inasmuch as $I \oplus J \in \mathscr{S}(R)$, this forces $I' = 0$, which is false.

(c) $\Rightarrow$ (b): Given any central idempotent $e \in Q$, we see that $I = R \cap eQ$ and $J = R \cap (1 - e)Q$ are two-sided ideals of R such that $I \cap J = 0$. Clearly $I_R \leqq_e (eQ)_R$ and $J_R \leqq_e [(1 - e)Q]_R$, whence $(I \oplus J)_R \leqq_e Q_R$ and consequently $I \oplus J \in \mathscr{S}(R)$. According to (c), either $I = 0$ or $J = 0$, from which we infer that either $e = 0$ or $1 - e = 0$.

(b) $\Rightarrow$ (a): Let I,J be any nonzero two-sided ideals of Q. If H is the $\mathscr{S}$-closure of J_Q in Q_Q, then H is a two-sided ideal of Q and $H \in L^*(Q_Q)$. Inasmuch as Q_Q is injective, 1.9 says that H_Q is injective; hence $H = eQ$ for some idempotent $e \in Q$. According to 2.33, e must be a central idempotent. Since $H \neq 0$, it follows from (b) that $e = 1$, and consequently $H = Q$. Thus $J \in \mathscr{S}(Q)$. Since $I \neq 0$ and $Z_r(Q) = 0$, we now conclude that $IJ \neq 0$. Therefore Q is prime. □

In particular, it follows from 6.1 that if R is prime, then so is $S^\circ R$. The converse fails in general (Exercise 1), but if R is semiprime, then primeness of $S^\circ R$ does imply primeness of R (Exercise 2).

In order to study nonzero prime ideals of $S^\circ R$, we need some results about principal right ideals in $S^\circ R$. Since any principal right ideal I of $S^\circ R$ is a direct summand of $S^\circ R$, we see that any such I is a nonsingular injective right R-module. For this reason, and also to allow for later use, we phrase the following results in terms of nonsingular injective modules.

Lemma 6.2 Let A,B be nonsingular injective right R-modules.

(a) Any $C \in L^*(A)$ is a nonsingular injective right R-module and is a direct summand of A.

(b) If $f \in \operatorname{Hom}_R(A,B)$, then $\ker f \in L^*(A)$, $fA \in L^*(B)$, and $A \cong (\ker f) \oplus (fA)$.

Proof: (a) According to 2.4, C is a closed submodule of A; hence 1.9

says that C is injective. Then C is a direct summand of A, and C is nonsingular because A is nonsingular.

(b) Inasmuch as $A/(\ker f)$ is isomorphic to a submodule of the nonsingular module B, we see that $\ker f \in L^*(A)$. According to (a), $\ker f$ is thus a direct summand of A; hence $A \cong (\ker f) \oplus (fA)$. Now fA is injective and so is a direct summand of B, from which we conclude that $fA \in L^*(B)$. □

Definition A module A is said to be *subisomorphic* to a module B, written $A \lesssim B$, provided A is isomorphic to a submodule of B.

If A and B are vector spaces over a field F, then either $A \lesssim B$ or $B \lesssim A$, depending on whether $[A : F] \leqq [B : F]$ or $[B : F] \leqq [A : F]$. In general, however, this is far from true. For example, if F is a field, $R = F \times F$, $A = F \times 0$, and $B = 0 \times F$, then neither $A \lesssim B$ nor $B \lesssim A$. The most that can be said here is that there is a ring decomposition $R = R_1 \times R_2$ such that $A \lesssim B$ "on R_2" and $B \lesssim A$ "on R_1." More precisely, there is a central idempotent $e = (0,1)$ in R such that $Ae \lesssim Be$ and $B(1 - e) \lesssim A(1 - e)$.

By restricting attention to nonsingular injective right R-modules, we can prove a similar result of this form, except that the central idempotent comes from $S^\circ R$ rather than from R. This causes no problems, for if A is a nonsingular injective right R-module, then $A = S^\circ A$ and so A is also a right $S^\circ R$-module.

Theorem 6.3 Let $Z_r(R) = 0$, and let A,B be any nonsingular injective right R-modules. Then there exists a central idempotent $e \in S^\circ R$ such that $Ae \lesssim Be$ and $B(1 - e) \lesssim A(1 - e)$.

Proof: Set $Q = S^\circ R$, and let $\mathfrak{A}$ denote the collection of all triples (A',B',f'), where $A' \leqq A$, $B' \leqq B$, and $f'\colon A' \to B'$ is an isomorphism. We define a partial order on $\mathfrak{A}$ by setting $(A',B',f') \leqq (A'',B'',f'')$ whenever $A' \leqq A''$, $B' \leqq B''$, and f'' is an extension of f'. Now $\mathfrak{A}$ is nonempty because $(0,0,0) \in \mathfrak{A}$, and it is easily seen that every nonempty chain in $\mathfrak{A}$ has an upper bound in $\mathfrak{A}$; hence by Zorn's Lemma there exists a maximal element $(A',B',f') \in \mathfrak{A}$.

Inasmuch as A and B are injective, A contains an injective hull A'' for A', B contains an injective hull B'' for B', and f' extends to an isomorphism $f''\colon A'' \to B''$. Then $(A'',B'',f'') \in \mathfrak{A}$, and the maximality of (A',B',f') implies that $(A'',B'',f'') = (A',B',f')$. Thus A' and B' are injective; hence $A = A' \oplus C$ and $B = B' \oplus D$ for suitable C,D.

Now $C = S^\circ C$ is a right Q-submodule of A; hence the annihilator $H = \{q \in Q \mid Cq = 0\}$ is a two-sided ideal of Q. In view of 2.8, A is a nonsingular right Q-module, from which we infer that $H \in L^*(Q_Q)$. Now H_Q is injective; hence $H = eQ$ for some idempotent $e \in Q$, and 2.33 shows that e

is a central idempotent. Inasmuch as $Ce = 0$, we obtain $Ae = A'e \cong B'e \leqq Be$. Therefore $Ae \lesssim Be$.

We claim that $D(1 - e) = 0$. Suppose, on the contrary, that there is a nonzero element $x \in D(1 - e)$. According to 2.8, xQ is a nonsingular right Q-module; hence 3.12 says that $(xQ)_Q$ is projective. Thus $xQ \cong tQ$ for some idempotent $t \in Q$, and we note that $tQe \cong xQe = 0$. Since $x \neq 0$, we have $t \neq 0$, and consequently $t \notin eQ = H$. Then $yt \neq 0$ for some $y \in C$, and as with xQ we see that $(ytQ)_Q$ is projective. Inasmuch as there is an epimorphism of $xQ \cong tQ$ onto ytQ, it follows that there must exist a monomorphism $f : ytQ \to xQ$. Since $yt \in C$ and $x \in D$, we obtain

$$(A' \oplus ytQ, B' \oplus f(ytQ), f' \oplus f) \in \mathfrak{A}$$

which contradicts the maximality of (A', B', f').

Thus $D(1 - e) = 0$ as claimed, whence

$$B(1 - e) = B'(1 - e) \cong A'(1 - e) \leqq A(1 - e)$$

Therefore $B(1 - e) \lesssim A(1 - e)$.□

Definition Given any module A and any positive integer n, we write nA to denote the direct sum of n copies of A. Since nA is also the direct product of n copies of A, nA is usually denoted A^n. However, since we shall later be dealing with direct sums αA for infinite cardinals α—and αA is in general not isomorphic to A^α (Exercise 3)—we use the notation nA for the sake of consistency.

Lemma 6.4 Let $Z_r(R) = 0$, let I be a right ideal in the ring $Q = S^\circ R$, and let $x \in Q$. Then $x \in QI$ if and only if $xQ \lesssim nI$ for some positive integer n.

Proof: If $x \in QI$, then $x \in a_1 I + \cdots + a_n I$ for some elements $a_1, \ldots, a_n \in Q$. Left multiplication by each a_i defines a map of $I \to Q$, and together these maps induce a map $f\colon nI \to Q$ such that $x \in f(nI)$. Now xQ is a direct summand of Q_Q and so is projective; hence there is a map $g : xQ \to nI$ such that fg is the inclusion map $xQ \to f(nI)$. Then g is a monomorphism, whence $xQ \lesssim nI$.

Conversely, assume that $xQ \lesssim nI$ for some $n > 0$. Now xQ is a direct summand of Q_Q and so is injective; hence xQ must be isomorphic to a direct summand of nI. Composing the natural injections $I \to nI$ with the projection $nI \to xQ$, we obtain maps $f_1, \ldots, f_n : I \to xQ$ such that $f_1 I + \cdots + f_n I = xQ$. Inasmuch as Q_Q is injective, each f_i must be left multiplication by some $a_i \in Q$, from which we obtain $xQ = a_1 I + \cdots + a_n I \leqq QI$.□

Lemma 6.4 also holds under somewhat more general conditions, as in Exercise 4.

Theorem 6.5 Let $Z_r(R) = 0$, and let P be a proper two-sided ideal of $S^\circ R$.

Then the following conditions are equivalent:

(a) P is a prime ideal.

(b) For all central idempotents $e \in S^\circ R$, either $e \in P$ or $1 - e \in P$.

(c) The two-sided ideals in the ring $S^\circ R/P$ are linearly ordered under inclusion.

Proof: Set $Q = S^\circ R$.

(a) $\Rightarrow$ (b): Since e is central, $eQ(1 - e) = 0 \subseteq P$, whence either $e \in P$ or $1 - e \in P$.

(b) $\Rightarrow$ (c): If not, then Q/P must have two-sided ideals I/P and J/P such that $I/P \nleqq J/P$ and $J/P \nleqq I/P$. Choose elements $x \in I - J$ and $y \in J - I$. According to 6.3, there exists a central idempotent $e \in Q$ such that $xQe \lesssim yQe$ and $yQ(1 - e) \lesssim xQ(1 - e)$, that is, $xeQ \lesssim yeQ$ and $(y - ye)Q \lesssim (x - xe)Q$. If $e \in P$, then $yeQ \leqq P \leqq I$. Inasmuch as $(y - ye)Q \lesssim (x - xe)Q \leqq I$, we obtain $yQ \lesssim 2I$, whence 6.4 says that $y \in I$. This is false, and if $1 - e \in P$ we likewise obtain $x \in J$, which is also false. Therefore neither e nor $1 - e$ belongs to P, which contradicts (b).

(c) $\Rightarrow$ (a): Let I,J be two-sided ideals of Q which properly contain P. According to (c), either $I/P \leqq J/P$ or $J/P \leqq I/P$; hence we may assume that $I/P \leqq J/P$. Since Q is regular, we must have $I^2 = I > P$, and consequently $IJ \geqq I^2 > P$. Therefore P is prime.□

Corollary 6.6 Let $Z_r(R) = 0$, and let P be any prime ideal in $S^\circ R$. If K is any proper two-sided ideal of $S^\circ R$ which contains P, then K is a prime ideal.

Proof: Condition (b) of 6.5 carries over from P to K.□

Definition A *minimal prime ideal* in a ring R is a prime ideal which does not properly contain any other prime ideal. For example, if R is an integral domain then 0 is a minimal prime ideal of R. Every prime ideal of R must contain at least one minimal prime ideal (Exercise 5), but in general a given prime ideal may contain many minimal prime ideals (Exercise 6). However, this cannot happen in $S^\circ R$, as the following corollary of 6.5 shows.

Corollary 6.7 Let $Z_r(R) = 0$, and let P be any prime ideal in $S^\circ R$.

(a) There is exactly one maximal two-sided ideal in $S^\circ R$ which contains P.

(b) There is exactly one minimal prime ideal in $S^\circ R$ which is contained in P.

Proof: Set $Q = S^\circ R$.

(a) According to 6.5, the two-sided ideals in Q/P are linearly ordered, from which we see that Q/P has exactly one maximal two-sided ideal.

(b) Let X be the collection of all central idempotents in Q which belong to P, and let M be the two-sided ideal of Q generated by X. If e is any central idempotent in Q, then 6.5 says that either $e \in X$ or $1 - e \in X$, whence either $e \in M$ or $1 - e \in M$. Since M is contained in P and thus is proper, 6.5 now shows that M is a prime ideal in Q.

It remains to show that M is a minimal prime ideal, and that it is the only minimal prime ideal contained in P. Both of these properties will follow if we show that every prime ideal K which is contained in P must contain M.

Given any $e \in X$, we see from 6.5 that either $e \in K$ or $1 - e \in K$. If $1 - e \in K$, then e and $1 - e$ both belong to P, which is impossible. Thus $e \in K$. Therefore $X \subseteq K$, and consequently $M \subseteq K$.□

Another way to phrase the results of 6.7 is to say that the partially ordered set $\mathscr{S}$ of all prime ideals in $S^\circ R$ is a disjoint union of intervals. To see this, let $\{H_\alpha\}$ be the set of all minimal prime ideals in $S^\circ R$, and for each α let K_α be the unique maximal two-sided ideal of $S^\circ R$ which contains H_α. The interval $[H_\alpha, K_\alpha]$ in $\mathscr{S}$ is just the set $\{P \in \mathscr{S} \mid H_\alpha \leqq P \leqq K_\alpha\}$. (By 6.6, we see that $[H_\alpha, K_\alpha]$ is also the set of all two-sided ideals of $S^\circ R$ which lie between H_α and K_α.) According to 6.7, $\mathscr{S}$ is the disjoint union of the intervals $[H_\alpha, K_\alpha]$.

Exercises

1. Find an example of a right nonsingular ring R such that $S^\circ R$ is prime but R is not prime.
2. Let $Z_r(R) = 0$ and assume that R is semiprime. Prove that $S^\circ R$ is prime if and only if R is prime.
3. Let F be a field, and let α be any infinite cardinal. If A is the direct sum of α copies of F and B is the direct product of α copies of F, prove that $A \not\cong B$.
4. Let Q be a regular ring, $x \in Q$, I a right ideal of Q. Prove that $x \in QI$ if and only if $xQ \lesssim nI$ for some $n > 0$.
5. Prove that every prime ideal in a ring R must contain at least one minimal prime ideal.
6. Let F be a field, $T = F[x_1, x_2, \ldots]$, P the ideal of T generated by the x_n. Set $y_n = x_1 x_3 x_5 \cdots x_{2n-1} x_{2n}$ for $n = 1, 2, \ldots$, let I be the ideal of T generated by the y_n, and set $R = T/I$. Prove that P/I is a prime ideal of R which contains infinitely many minimal prime ideals.
7. Use 4.16 to prove that (b) ⇒ (c) in 6.1.
8. Let $Z_r(R) = 0$. Prove that $S^\circ R$ is prime if and only if for any nonzero two-sided ideal I of R whose left annihilator J is nonzero, we have $I \cap J \neq 0$.
9. If A and B are nonsingular injective right R-modules, show that $\mathrm{Hom}_R(A,B) = 0$ if and only if $\mathrm{Hom}_R(B,A) = 0$.
10. If $\{A_\alpha\}$ is a collection of nonsingular injective right R-modules such that $\mathrm{Hom}_R(A_\alpha, A_\beta) = 0$ whenever $\alpha \neq \beta$, prove that the endomorphism ring of $E(\bigoplus A_\alpha)$ is isomorphic to $\Pi \mathrm{End}_R(A_\alpha)$.

11. Suppose that for all right R-modules A,B there is a central idempotent $e \in R$ such that $Ae \lesssim Be$ and $B(1-e) \lesssim A(1-e)$. Prove that R is semisimple.
12. Prove that R is simple artinian if and only if for all right R-modules A,B, either $A \lesssim B$ or $B \lesssim A$.
13. Let $Z_r(R) = 0$, and let P be a prime ideal in R. Prove that either $P \in \mathscr{S}(R)$ or $P \in L^*(R_R)$, but not both.
14. Let R be the ring of all lower triangular 2×2 matrices over a field F, and recall from 4.3 that $Z_r(R) = Z_l(R) = 0$. Show that R has a prime ideal P such that $P_R \leqq_e R_R$ while ${}_RP \in L^*({}_RR)$.
15. Let R be regular and right self-injective. Prove that R is simple if and only if R has exactly one prime ideal.
16. Let $Z_r(R) = 0$, and let B be the set of all central idempotents in $S°R$, which by Exercise 2.D.18 forms a complete Boolean algebra. Prove that the rule $P \mapsto P \cap B$ defines a bijection between the minimal prime ideals of $S°R$ and the maximal ideals of B.
17. With notation as in Exercise 16, let $\mathscr{K}$ denote the collection of all minimal prime ideals of $S°R$, and let $\mathscr{M}$ denote the collection of all maximal ideals of B. There are standard "hull-kernel" topologies for $\mathscr{K}$ and $\mathscr{M}$: The open sets in $\mathscr{K}$ are those of the form $\{P \in \mathscr{K} \mid X \nsubseteq P\}$ (for any $X \subseteq S°R$), and the open sets in $\mathscr{M}$ are those of the form $\{M \in \mathscr{M} \mid Y \nsubseteq M\}$ (for any $Y \subseteq B$). Prove that the rule $P \mapsto P \cap B$ defines a homeomorphism of $\mathscr{K}$ onto $\mathscr{M}$. Conclude that $\mathscr{K}$ is compact, Hausdorff, and extremally disconnected.
18. Let $Z_r(R) = 0$, $n > 0$, T the ring of all $n \times n$ matrices over R. Prove that the space of all minimal prime ideals of $S°T$ (as defined in Exercise 17) is homeomorphic to the space of all minimal prime ideals of $S°R$.
19. Let $\{R_\lambda \mid \lambda \in X\}$ be a collection of prime rings, and set $R = \Pi R_\lambda$. Prove that there is a bijection between the ultrafilters on X and the minimal prime ideals of R.
20. Let $Z_r(R) = 0$, and let $\{e_\lambda\}$ be any collection of orthogonal central idempotents in $Q = S°R$. Prove that $Q/(\cap(1-e_\lambda)Q) \cong \Pi(Q/(1-e_\lambda)Q)$.
21. Let $Z_r(R) = 0$, and let $\{P_\lambda\}$ be a nonempty collection of prime ideals in $Q = S°R$. If there exist orthogonal central idempotents $\{e_\lambda\} \subset Q$ such that $e_\lambda \notin P_\lambda$ for all λ, prove that $Q/(\cap P_\lambda) \cong \Pi(Q/P_\lambda)$.

B. Directly Finite Modules

Definition A module A is *directly finite* (*von Neumann finite, Dedekind finite*) provided A is not isomorphic to any proper direct summand of itself. If A is not directly finite, then A is said to be *directly infinite*.

Obviously all finite modules are directly finite, as are all modules with finite length (Exercise 1). Any indecomposable module is directly finite. A vector space is a directly finite module if and only if it is finite-dimensional. More generally, all finite-dimensional modules are directly finite (Exercise 2), but directly finite modules need not be finite-dimensional (Exercise 3). Any direct summand of a directly finite module is directly finite (Exercise 4), but submodules and factor modules of directly finite modules need not be directly finite (Exercises 5 and 6).

Lemma 6.8 If A is an injective module and B is a directly finite module such that $A \lesssim B$, then A is directly finite.

Proof: This follows immediately from the observation that A is isomorphic to a direct summand of B.□

Lemma 6.9 A right R-module A is directly finite if and only if for all $f,g \in \operatorname{End}_R(A)$, $fg = 1$ implies $gf = 1$.

Proof: First assume that A is directly finite, and suppose that $f,g \in \operatorname{End}_R(A)$ with $fg = 1$. Since $1 - gf$ is an idempotent endomorphism of A such that $(1 - gf)g = 0$, we obtain $A = gA \oplus (1 - gf)A$. Then gA is a direct summand of A which is isomorphic to A; hence $gA = A$ and so $(1 - gf)A = 0$. Thus $gf = 1$.

Conversely, if A is not directly finite, we must have $A = B \oplus C$ with $A \cong B$ and $C \neq 0$. There is an isomorphism $g : A \to B$, and we can define a map $f: A \to A$ such that $f|_B = g^{-1}$ and $f|_C = 0$. Then $fg = 1$, but since $gfC = 0$ we see that $gf \neq 1$.□

In particular, 6.9 shows that R_R is directly finite if and only if $xy = 1$ implies $yx = 1$ in R, that is, if and only if all one-sided inverses in R are two-sided. Since this condition is left-right symmetric, we see that R_R is directly finite if and only if ${}_RR$ is directly finite, in which case we say that R is a *directly finite ring*. For example, Exercise 2.D.20 says that any right and left self-injective ring is directly finite. Also, 6.9 says that a module A_R is directly finite if and only if $\operatorname{End}_R(A)$ is a directly finite ring. Note that any subring of a directly finite ring is directly finite, and that any direct product of directly finite rings is directly finite.

Proposition 6.10 For any injective module A, the following conditions are equivalent:

(a) A is directly infinite.
(b) A has a nonzero submodule M such that $2M \cong M$.
(c) A has a nonzero direct summand B such that $2B \cong B$.

Proof: (b) ⇒ (c): Since A is injective, it must contain an injective hull B for M, and we note that B is a nonzero direct summand of A. Since $2M \cong M$, we must have $2B \cong B$.

(c) ⇒ (a): Since B is an injective directly infinite module, it follows from 6.8 that A must be directly infinite.

(a) ⇒ (b): We must have $A = C \oplus D$ with $A \cong C$ and $D \neq 0$. Choosing an isomorphism $f\colon A \to C$, we claim that $\{D, fD, f^2D, \ldots\}$ is an independent sequence of submodules of A. Obviously $\{D\}$ is independent. If $\{D, fD, \ldots, f^nD\}$ is independent for some $n > 0$, then since f is an isomorphism we see that $\{fD, f^2D, \ldots, f^{n+1}D\}$ must be independent. Inasmuch as

$$D \cap (fD \oplus f^2D \oplus \cdots \oplus f^{n+1}D) \leqq D \cap fA = D \cap C = 0$$

it follows that $\{D, fD, \ldots, f^{n+1}D\}$ is independent. Thus the induction works, and consequently $\{D, fD, f^2D, \ldots\}$ is independent, as claimed. As a result, A has a nonzero submodule $M = D \oplus fD \oplus f^2D \oplus \cdots$. Inasmuch as $f^nD \cong D$ for all n, we conclude that $M \cong 2M$. □

Example 6.11 There exists a right nonsingular ring R which has nonsingular right modules A, B such that A and B are directly finite while $A \oplus B$ is directly infinite.

Proof: Set $R = \mathbf{Z}$, which is a nonsingular ring. Let V be a vector space over $\mathbf{Q}$ with basis $\{a_k, b_k, x_k, y_k \mid k = 1,2,\ldots\}$, and let $q_1, q_2, \ldots$ be distinct prime numbers all strictly greater than 5. Define R-submodules $A, B \leqq V$ as follows:

$$A \text{ is generated by } a_k/5^n,\ x_k/q_k^n,\ (a_k + x_k)/3\ (n,k = 1,2,\ldots)$$

$$B \text{ is generated by } b_k/5^n,\ y_k/q_k^n,\ (b_k + y_k)/2\ (n,k = 1,2,\ldots)$$

Note that A and B are nonsingular R-modules, and that $A \cap B = 0$.

Set $c_k = a_k - b_k$ and $d_k = b_k3 - a_k2$ for all k, and define additional R-submodules $C, D \leqq V$ as follows:

$$C \text{ is generated by } c_k/5^n\ (n,k = 1,2,\ldots)$$

$$D \text{ is generated by } d_k/5^n,\ x_k/q_k^n,\ y_k/q_k^n$$
$$(d_k + x_k)/3,\ (d_k + y_k)/2\ (n,k = 1,2,\ldots)$$

Observing that $\{c_k, d_k, x_k, y_k \mid k = 1,2,\ldots\}$ is a basis for V, we see that $C \cap D = 0$. Inasmuch as

$$a_k = c_k3 + d_k \qquad b_k = c_k2 + d_k$$

$$(a_k + x_k)/3 = c_k + (d_k + x_k)/3 \qquad (b_k + y_k)/2 = c_k + (d_k + y_k)/2$$

$$(d_k + x_k)/3 = b_k - a_k + (a_k + x_k)/3 \qquad (d_k + y_k)/2 = b_k - a_k + (b_k + y_k)/2$$

for all k, we infer that $A \oplus B = C \oplus D$.

Now $C = C_1 \oplus C_2 \oplus \cdots$, where each C_k is generated by $\{c_k/5^n \mid n = 1,2,\ldots\}$. Since these C_k are all isomorphic, we obtain $C \cong 2C$, whence C is directly infinite. Inasmuch as C is a direct summand of $A \oplus B$, it follows that $A \oplus B$ is directly infinite.

Let A' denote the submodule of A generated by $\{a_k/5^n \mid n,k = 1,2,\ldots\}$. Inasmuch as all $q_k > 5$, we infer that $A' = \bigcap_{n=1}^{\infty} A5^n$, whence A' is a fully invariant submodule of A. For each $k = 1,2,\ldots$, let A_k denote the submodule of A generated by $\{x_k/q_k^n \mid n = 1,2,\ldots\}$. We infer that $A_k = \bigcap_{n=1}^{\infty} Aq_k^n$, so that A_k is a fully invariant submodule of A.

We claim that if $p_k \in A_k$ for $k = 1,2,\ldots,s$ such that $p_1 + \cdots + p_s \in A3$, then each $p_k \in A_k3$. There must exist integers α_{nk}, β_{nk}, γ_k $(n,k = 1,2,\ldots)$, all but finitely many of which are zero, such that

$$\begin{aligned} p_1 + \cdots + p_s &= [\sum_{n,k} a_k(\alpha_{nk}/5^n) + \sum_{n,k} x_k(\beta_{nk}/q_k^n) + \sum_k (a_k + x_k)(\gamma_k/3)]3 \\ &= \sum_k a_k[\gamma_k + 3\sum_n (\alpha_{nk}/5^n)] + \sum_k x_k[\gamma_k + 3\sum_n (\beta_{nk}/q_k^n)] \end{aligned}$$

As a result, we find that for $k = 1,\ldots,s$, $\gamma_k + 3\sum_n (\alpha_{nk}/5^n) = 0$ and $p_k = x_k[\gamma_k + 3\sum_n (\beta_{nk}/q_k^n)]$. Since 3 and 5 are relatively prime, it follows from the first equation that each γ_k is divisible by 3, and then from the second equation that each $p_k \in A_k3$, as claimed.

Now consider any $f,g \in \mathrm{End}_R(A)$ such that $fg = 1$. Since $A',A_1,A_2,\ldots$ are fully invariant submodules of A, f and g restrict to endomorphisms of $A',A_1,A_2,\ldots$.

For $k = 1,2,\ldots$, we see that A_k is indecomposable and hence directly finite. Inasmuch as $fg = 1$ on A_k, we thus obtain $gf = 1$ on A_k (by 6.9); hence f restricts to an automorphism of A_k. Observing that $x_k \notin A_k3$, it now follows that $fx_k \notin A_k3$ as well.

We claim that $A' \cap (\ker f) = 0$. If not, then there exist integers $m_1,\ldots,m_s$ and $n(1),\ldots,n(s)$ such that $m_s \neq 0$ and

$$f[a_1(m_1/5^{n(1)}) + \cdots + a_s(m_s/5^{n(s)})] = 0$$

Multiplying through by a suitable power of 5, and dividing by the greatest common divisor of $m_1,\ldots,m_s$, we obtain $f(a_1t_1 + \cdots + a_st_s) = 0$ for some integers $t_1,\ldots,t_s$ whose greatest common divisor is 1. Inasmuch as each $a_k + x_k \in A3$, we now compute that

$$(fx_1)t_1 + \cdots + (fx_s)t_s = f[(a_1 + x_1)t_1 + \cdots + (a_s + x_s)t_s] \in A3$$

Now each $(fx_k)t_k \in A_k$; hence it follows from the claim above that each $(fx_k)t_k \in A_k3$. However, $fx_k \notin A_k3$, from which we conclude that each t_k is divisible by 3, which is impossible.

Therefore $A' \cap (\ker f) = 0$; hence f is injective on A' as well as on A_1, $A_2, \ldots$. Inasmuch as $A', A_1, A_2, \ldots$ are independent fully invariant submodules of A, it follows that f must be injective on $A' \oplus A_1 \oplus A_2 \oplus \cdots$. Observing that $A3 \leqq A' \oplus A_1 \oplus A_2 \oplus \cdots$, we conclude that f is a monomorphism. Since $f(1 - gf) = 0$, we obtain $gf = 1$, and consequently A is directly finite, by 6.9.

In exactly the same manner, it can be shown that B is directly finite.□

In spite of 6.11, it is possible to show that the direct sum of two directly finite nonsingular injective modules is directly finite, for which we need the following lemmas.

Lemma 6.12 Let A,B be nonsingular injective right R-modules, and let $C \in L^*(A \oplus B)$. Then there exist decompositions $A = A_1 \oplus A_2$ and $B = B_1 \oplus B_2$ such that $A_1 \oplus B_1 \cong C$ and $A_2 \oplus B_2 \cong (A \oplus B)/C$.

Proof: Let $p_1 : A \oplus B \to A$ and $p_2 : A \oplus B \to B$ denote the projection maps, and set $A_1 = C \cap A$, $B_1 = p_2C$. Now C is a nonsingular injective module by 6.2, and A_1 is the kernel of the restricted projection $p_2 : C \to B$; hence it follows from 6.2 that $C \cong A_1 \oplus B_1$. Then A_1 and B_1 are injective; hence we must have $A = A_1 \oplus A_2$ and $B = B_1 \oplus B_2$ for suitable A_2,B_2.

Observing that $B_1 = (1 - p_1)C \leqq C + A$, we see that $B \leqq C + A + B_2$, and thus $C + A + B_2 = A \oplus B$. We also have $A = A_1 + A_2 \leqq C + A_2$; hence $C + A = C + A_2$, and consequently $A \oplus B = C + A_2 + B_2$.

Now set $F = C \cap (A_2 + B_2)$. Inasmuch as $p_2C = B_1$ while $p_2(A_2 + B_2) = B_2$, we find that $p_2F \leqq B_1 \cap B_2 = 0$, and consequently $F \leqq A$. As a result, $F \leqq C \cap A = A_1$ and $p_1F = F$. We also have $p_1F \leqq p_1(A_2 + B_2) = A_2$, whence $F \leqq A_1 \cap A_2 = 0$.

Therefore $A \oplus B = C \oplus (A_2 \oplus B_2)$; hence $(A \oplus B)/C \cong A_2 \oplus B_2$.□

Lemma 6.13 Let $Z_r(R) = 0$, and let A,B,C,D be nonsingular injective right R-modules such that $A \oplus B \cong C \oplus D$. Then there exists a central idempotent $e \in S^\circ R$ such that $Ae \lesssim Ce$ and $B(1 - e) \lesssim D(1 - e)$.

Proof: In view of 6.12, there exist decompositions $A = A_1 \oplus A_2$ and $B = B_1 \oplus B_2$ such that $A_1 \oplus B_1 \cong C$ and $A_2 \oplus B_2 \cong D$. According to 6.3, there is a central idempotent $e \in S^\circ R$ such that $A_2e \lesssim B_1e$ and $B_1(1 - e) \lesssim A_2(1 - e)$. Thus

$$Ae = A_1e \oplus A_2e \lesssim A_1e \oplus B_1e \cong Ce$$

and similarly

$$B(1 - e) = B_1(1 - e) \oplus B_2(1 - e) \lesssim A_2(1 - e) \oplus B_2(1 - e) \cong D(1 - e)$$□

Theorem 6.14 Let $Z_r(R) = 0$, and let A,B be nonsingular injective right R-modules. If A and B are both directly finite, then $A \oplus B$ is directly finite.

Proof: Suppose that $C \in L^*(A \oplus B)$ with $C \cong 2C$. Using 6.12, we obtain $C \cong A_1 \oplus B_1$ for some $A_1 \in L^*(A)$ and some $B_1 \in L^*(B)$. Then $C \oplus C \cong A_1 \oplus B_1$; hence by 6.13 there is a central idempotent $e \in S^\circ R$ such that $Ce \lesssim A_1 e$ and $C(1-e) \lesssim B_1(1-e)$. Now $Ce \lesssim A_1 \leqq A$; hence 6.8 says that Ce is directly finite. However, $Ce \cong 2(Ce)$; hence we see from 6.10 that $Ce = 0$. We likewise obtain $C(1-e) = 0$, whence $C = 0$. Now 6.10 shows that $A \oplus B$ is directly finite.□

Corollary 6.15 Let $Z_r(R) = 0$, $Q = S^\circ R$. Then the set $F = \{x \in Q \mid (xQ)_R$ is directly finite$\}$ is a two-sided ideal of Q.

Proof: Recall that all principal right ideals of Q are also nonsingular injective right R-modules.

Obviously $0 \in F$.

Given $x \in F$ and $y \in Q$, we have xQ directly finite and $xyQ \leqq xQ$; hence 6.8 says that xyQ is directly finite. Thus $xy \in F$. Left multiplication by y defines an epimorphism of xQ onto yxQ; hence we see from 6.2 that $yxQ \lesssim xQ$. Again, 6.8 shows that yxQ is directly finite, and so $yx \in F$.

Given $x,y \in F$, 6.14 shows that $xQ \oplus yQ$ is directly finite. We have an epimorphism of $xQ \oplus yQ$ onto $xQ + yQ$; hence 6.2 shows that $xQ + yQ \lesssim xQ \oplus yQ$, and consequently $(x-y)Q \lesssim xQ \oplus yQ$. Then $(x-y)Q$ is directly finite by 6.8, and thus $x - y \in F$.□

Theorem 6.16 Let $Z_r(R) = 0$, and let A,B,C be nonsingular injective right R-modules. If A is directly finite and $A \oplus B \cong A \oplus C$, then $B \cong C$.

Proof: In view of 6.12, there must exist decompositions $A = A_1 \oplus A_2$ and $B = B_1 \oplus B_2$ such that $A_1 \oplus B_1 \cong A$ and $A_2 \oplus B_2 \cong C$. According to 6.3, there is a central idempotent $e \in S^\circ R$ such that $A_2 e \lesssim B_1 e$ and $B_1(1-e) \lesssim A_2(1-e)$. If $A_2 e$ is isomorphic to a proper direct summand of $B_1 e$, then $Ae = A_1 e \oplus A_2 e$ is isomorphic to a proper direct summand of $A_1 e \oplus B_1 e \cong Ae$. In this case Ae is directly infinite, which contradicts 6.8. Thus we must have $A_2 e \cong B_1 e$, and similarly $B_1(1-e) \cong A_2(1-e)$. Therefore $A_2 \cong B_1$, whence $C \cong A_2 \oplus B_2 \cong B_1 \oplus B_2 = B$.□

Corollary 6.17 Let $Z_r(R) = 0$, and let A,B,C be nonsingular injective right R-modules. If A is directly finite and $A \oplus B \lesssim A \oplus C$, then $B \lesssim C$.

Proof: Since $A \oplus B$ is injective, we must have $A \oplus C \cong A \oplus B \oplus D$ for some D. According to 6.16, $C \cong B \oplus D$, and consequently $B \lesssim C$.□

Theorem 6.16 is actually one step in the proof of a more general cancellation theorem for quasi-injective modules, for which we need two intermediate results.

Proposition 6.18 Suppose that we are given module decompositions $M = A \oplus B = \oplus C_i$. If A is quasi-injective, then there exist submodules $C_i' \leqq C_i$ for all i such that $M = A \oplus (\oplus C_i')$.

Proof: Let X denote the collection of all submodules $C \leqq M$ such that (a) $\oplus(B \cap C_i) \leqq C$; (b) $A \cap C = 0$; (c) $C = \oplus(C \cap C_i)$. Since $\oplus(B \cap C_i) \in X$, X is nonempty. Observing that the union of any nonempty chain in X yields a member of X, we see from Zorn's Lemma that there is a maximal element $C \in X$. Inasmuch as $M = \oplus C_i$ and $C = \oplus(C \cap C_i)$, we note that $M/C = \oplus[(C_i + C)/C]$.

We claim that $[(A + C)/C] \cap [(C_j + C)/C] \leqq_e (C_j + C)/C$ for any index j. Thus consider any submodule $D/C \leqq (C_j + C)/C$ for which $[(A + C)/C] \cap [D/C] = 0$. Then $(A + C) \cap D = C$, and consequently $A \cap D \leqq A \cap C = 0$. In addition, $D = C_j^* + C$ for some $C_j^* \leqq C_j$; hence

$$D = C_j^* + [\oplus(C \cap C_i)] = [C_j^* + (C \cap C_j)] \oplus [\bigoplus_{i \neq j} (C \cap C_i)]$$

from which we infer that $D = \oplus(D \cap C_i)$. Since $\oplus(B \cap C_i) \leqq C \leqq D$, we now have $D \in X$; hence it follows from the maximality of C that $D = C$ and so $D/C = 0$. Thus the claim is proved.

As a result, $\oplus([(A + C)/C] \cap [(C_i + C)/C]) \leqq_e \oplus[(C_i + C)/C] = M/C$, and consequently $(A + C)/C \leqq_e M/C$. Thus $E[(A + C)/C] = E(M/C)$. Since $A \cap C = 0$, $(A + C)/C$ is isomorphic to A and so is quasi-injective. Therefore $(A + C)/C$ is a fully invariant submodule of $E(M/C)$, by 2.13.

Given any index j, $B \cap C_j$ is the kernel of the projection $C_j \to M = A \oplus B \to A$; hence $C_j/(B \cap C_j)$ is isomorphic to a submodule of A. In addition, $(A + C)/C \cong A$; hence $C_j/(B \cap C_j) \cong D_j$ for some $D_j \leqq (A + C)/C$. On the other hand, $B \cap C_j \leqq C$ and so $B \cap C_j \leqq C \cap C_j$; hence there is an epimorphism of $C_j/(B \cap C_j)$ onto $C_j/(C \cap C_j) \cong (C_j + C)/C$. Thus there is an epimorphism $f : D_j \to (C_j + C)/C$, and f extends to an endomorphism g of $E(M/C)$. Inasmuch as $(A + C)/C$ is fully invariant in $E(M/C)$, we obtain

$$(C_j + C)/C = fD_j = gD_j \leqq g[(A + C)/C] \leqq (A + C)/C$$

Now $M/C = \oplus[(C_i + C)/C] \leqq (A + C)/C$, whence $M = A + C = A \oplus C$. Since $C \in X$, we have $C = \oplus C_i'$, where $C_i' = C \cap C_i \leqq C_i$. Therefore $M = A \oplus (\oplus C_i')$. □

Lemma 6.19 Let A be a directly finite quasi-injective right R-module. If $A = B \oplus B' = C \oplus C'$ with $B \cong C$, then $B' \cong C'$.

Proof: Since A is directly finite, 6.9 shows that the ring $Q = \mathrm{End}_R(A)$ is directly finite. We claim that the ring $\bar{Q} = Q/J(Q)$ is directly finite as well. Thus consider any $a,b \in Q$ for which $\overline{ab} = 1$ in $\bar{Q}$. Then $ab - 1 \in J(Q)$, hence $ab = 1 + (ab - 1)$ is invertible in Q. Now $abc = 1$ for some $c \in Q$, and since Q is directly finite we have $bca = 1$ as well. Consequently, $\overline{bca} = 1$ in $\bar{Q}$, whence $\overline{bc} = (\overline{bc})(\overline{ab}) = \bar{b}$, and so $\overline{ba} = 1$. Thus $\bar{Q}$ is indeed directly finite.

Because of the given decompositions of A, there exist idempotents $e,f \in Q$ such that $eA = B'$, $(1 - e)A = B$, $fA = C'$, and $(1 - f)A = C$. Inasmuch as $(1 - e)A \cong (1 - f)A$, there exist elements $x \in (1 - e)Q(1 - f)$ and $y \in (1 - f)Q(1 - e)$ such that $xy = 1 - e$ and $yx = 1 - f$. Then $\overline{xy} = 1 - \bar{e}$ and $\overline{yx} = 1 - \bar{f}$; hence $(1 - \bar{e})\bar{Q} \cong (1 - \bar{f})\bar{Q}$, and consequently

$$(1 - \bar{e})\bar{Q} \oplus \overline{eQ} = \bar{Q} = (1 - \bar{f})\bar{Q} \oplus \overline{fQ} \cong (1 - \bar{e})\bar{Q} \oplus \overline{fQ}$$

According to 2.21, $\bar{Q}$ is a regular, right self-injective ring, hence $(1 - \bar{e})\bar{Q}$, $\overline{eQ}$, and $\overline{fQ}$ are nonsingular injective right $\bar{Q}$-modules. Since $\bar{Q}$ is directly finite, so is $(1 - \bar{e})\bar{Q}$; hence 6.16 shows that $\overline{eQ} \cong \overline{fQ}$.

Thus there must exist elements $\bar{z} \in \overline{eQf}$ and $\bar{w} \in \overline{fQe}$ such that $\overline{zw} = \bar{e}$ and $\overline{wz} = \bar{f}$. Replacing z by ezf and w by fwe, we may assume in addition that $z \in eQf$ and $w \in fQe$. Now $zw - e \in J(Q)$; hence $(zw + 1 - e)p = 1$ for some $p \in Q$. Multiplying this equation on the left by $1 - e$, we obtain $(1 - e)p = 1 - e$, and consequently $zwp + 1 - e = 1$. Thus $zwp = e$, and by symmetry $qwz = f$ for some $q \in Q$.

Inasmuch as $z \in eQf$ and $qwz = f$, the rule $\phi(a) = za$ defines a monomorphism $\phi : fA \to eA$. In addition, $eA = zwpA \leqq \phi(fA)$, whence ϕ is surjective. Therefore $fA \cong eA$, that is, $C' \cong B'$. □

Theorem 6.20 Let A be a directly finite quasi-injective right R-module. If B,C are any right R-modules such that $A \oplus B \cong A \oplus C$, then $B \cong C$.

Proof: Setting $M = A \oplus B$, we have $M = A_0 \oplus C_0$ for some $A_0 \cong A$ and some $C_0 \cong C$. According to 6.18, there exist submodules $A' \leqq A_0$ and $C' \leqq C_0$ such that $M = A \oplus A' \oplus C'$. Since $A' \leqq A_0 \leqq M$ and A' is a direct summand of M, it must also be a direct summand of A_0. Thus $A_0 = A' \oplus A''$ for some A'', and likewise $C_0 = C' \oplus C''$ for some C''.

Now $M = A \oplus A' \oplus C'$ and $M = A_0 \oplus C_0 = A' \oplus C' \oplus A'' \oplus C''$; hence $A \cong M/(A' \oplus C') \cong A'' \oplus C''$. On the other hand, $A \cong A_0 = A' \oplus A''$, hence we conclude from 6.19 that $A' \cong C''$. Therefore

$$B \cong M/A \cong A' \oplus C' \cong C'' \oplus C' = C_0 \cong C$$ □

Corollary 6.21 If $A_1, \ldots, A_n$ are directly finite quasi-injective right R-modules, then $A_1 \oplus \cdots \oplus A_n$ is directly finite.

Proof: Suppose that $A_1 \oplus \cdots \oplus A_n = B \oplus C$ with $A_1 \oplus \cdots \oplus A_n \cong$

B. Then $A_1 \oplus \cdots \oplus A_n \oplus C \cong A_1 \oplus \cdots \oplus A_n \oplus 0$; hence it follows from 6.20 that $C = 0$. Therefore $B = A_1 \oplus \cdots \oplus A_n$.□

In the situation described in 6.21, $A_1 \oplus \cdots \oplus A_n$ need not be quasi-injective. For example, let $A_1 = \mathbf{Q}$ and $A_2 = \mathbf{Z}/2\mathbf{Z}$, which are quasi-injective $\mathbf{Z}$-modules. Since A_1 and A_2 are indecomposable, they are also directly finite. However, $A_1 \oplus A_2$ is not quasi-injective, as shown in the discussion following 2.15.

Exercises

1. Show that any module with finite length is a directly finite module.
2. Prove that any finite-dimensional module is directly finite.
3. Find an example of a directly finite module which is not finite-dimensional.
4. Show that any direct summand of a directly finite module is directly finite.
5. Let F be a field, V an infinite-dimensional vector space over F, $R = \begin{pmatrix} F & 0 \\ V & F \end{pmatrix}$. Show that R_R is directly finite, but that R has a right ideal which is not directly finite.
6. With A,B as in Exercise 3.A.4, show that A is directly finite but that A/B is not.
7. Let F be a field, $F_n = F$ for $n = 1,2,\ldots,$ $R = \Pi F_n$. Find an example of a nonsingular injective R-module which is directly finite but not finitely generated.
8. If $n > 0$, prove that a $(\mathbf{Z}/n\mathbf{Z})$-module is directly finite if and only if it is finitely generated.
9. Find an example of a right nonsingular ring R which has a nonsingular injective right module A such that A is finitely generated but not directly finite.
10. If R is an abelian regular ring (Section 2.D), prove that R is directly finite.
11. Prove that $\begin{pmatrix} A & 0 \\ B & C \end{pmatrix}$ is directly finite if and only if A and C are directly finite.
12. Prove that R is directly finite if and only if $R/J(R)$ is directly finite.
13. Prove that any ring Morita-equivalent to a commutative ring must be directly finite.
14. If R is an integral domain which is not right Ore, prove that R is directly finite but that $S^\circ R$ is directly infinite.
15. Let R be right self-injective, let $n > 0$, and let T be the ring of all $n \times n$ matrices over R. Prove that R is directly finite if and only if T is directly finite.
16. If A and B are directly finite modules such that $\mathrm{Hom}_R(A,B) = 0$ and $\mathrm{Hom}_R(B,A) = 0$, show that $A \oplus B$ is directly finite.
17. Let $\{A_\lambda\}$ be a collection of nonsingular injective modules such that $\mathrm{Hom}_R(A_\lambda,A_\gamma) = 0$ whenever $\lambda \neq \gamma$. Prove that $E(\bigoplus A_\lambda)$ is directly finite if and only if each A_λ is directly finite.
18. A ring R is *unit-regular* if for any $x \in R$, there exists a unit $u \in R$ such that $xux = x$. Prove that a regular, right self-injective ring R is unit-regular if and only if it is directly finite.

19. Prove 6.14 by using 6.16.

20. Let $Z_r(R) = 0$, and let A be a nonsingular injective right R-module. Prove that there is a central idempotent $e \in S^{\circ}R$ such that Ae is directly finite and $A(1-e) \cong 2A(1-e)$.

21. Let $Z_r(R) = 0$, and let A,B be nonsingular injective right R-modules. If $nA \lesssim nB$ for some $n > 0$, prove that $A \lesssim B$.

22. Let $Z_r(R) = 0$, let A,B be nonsingular injective right R-modules, and let X be a nonempty chain in $L^*(A)$. If each member of X is directly finite and subisomorphic to B, prove that $E(\cup X) \lesssim B$.

23. Let $Z_r(R) = 0$, and let A,B be nonsingular injective right R-modules. If B is directly finite and $nA \lesssim B$ for all $n > 0$, prove that $A = 0$.

C. Directly Infinite Modules

Definition Let A be a module, and let α be any positive cardinal number (finite or infinite). We write αA to stand for the direct *sum* of α copies of A, and we say that A has *multiplicity* α if $\alpha A \lesssim A$. Obviously every module has multiplicity 1, and the zero module has multiplicity α for all α. If V is a nonzero finite-dimensional vector space over a field F, then 1 is the only multiplicity of V (Exercise 1). If V is infinite-dimensional, then V has multiplicity α for all $\alpha \leqq [V:F]$ (Exercise 2). Note that if a module A has multiplicity α, then A also has multiplicity β for all $\beta \leqq \alpha$.

For an injective module A with multiplicity α, it follows from the assumption $\alpha A \lesssim A$ that $E(\alpha A) \lesssim A$ as well. Since $\alpha \geqq 1$, we also have $A \lesssim E(\alpha A)$; hence 1.13 shows that $A \cong E(\alpha A)$. Thus, *an injective module A has multiplicity α if and only if $E(\alpha A) \cong A$.*

By analogy with the vector space situation, we shall use the concept of multiplicity to develop a type of dimension. We first note that only infinite multiplicities need to be used, since any module with multiplicity $\alpha \geqq 2$ also has multiplicity $\aleph_0$ (Exercise 3). Secondly, we note that as long as $A \neq 0$, there is an upper bound on the possible multiplicities of A. For if $\alpha A \lesssim A$, then A contains an independent family consisting of α nonzero submodules, which is possible only for $\alpha \leqq \text{card}(A)$. Unfortunately, there is no guarantee that A must have a largest multiplicity; hence we must work instead with the smallest non-multiplicity.

Definition Let A be any module. If $A = 0$, define $\mu(A) = 0$. If $A \neq 0$, define $\mu(A)$ to be the smallest *infinite* cardinal α such that A does *not* have multiplicity α. For example, it follows from Exercises 1 and 2 that a vector space V is finite-dimensional if and only if $\mu(V) \leqq \aleph_0$. If V is infinite-dimensional, then Exercise 2 shows that $\mu(V)$ is the successor of the dimension of V.

By analogy with this example, we might expect a module A to be directly finite if and only if $\mu(A) \leqq \aleph_0$. However, there exist directly finite modules A with $\mu(A) > \aleph_0$ (Exercise 4), and there exist directly infinite modules A with $\mu(A) \leqq \aleph_0$. For example, let F be a field, $R = F \times F$, V an infinite-dimensional vector space over F, $A = F \times V$. Because the left-hand factor of A is directly finite, we see that $\mu(A) = \aleph_0$. However, the right-hand factor of A is directly infinite, hence A is directly infinite.

In order to avoid this and other problems in studying nonsingular injective modules (see Exercises 5 and 6), we restrict attention to the case when $S^{\circ}R$ is an indecomposable ring, which by 6.1 is equivalent to the assumption that $S^{\circ}R$ is a prime ring.

Theorem 6.22 Let $Z_r(R) = 0$, and assume that $S^{\circ}R$ is prime. If A,B are nonsingular injective right R-modules, then either $A \lesssim B$ or $B \lesssim A$.

Proof: Since the only central idempotents in $S^{\circ}R$ are 0 and 1, this follows immediately from 6.3. □

Proposition 6.23 Let $Z_r(R) = 0$, and assume that $S^{\circ}R$ is prime. If A,B are nonsingular injective right R-modules with $A \lesssim B$, then $\mu(A) \leqq \mu(B)$.

Proof: Since this is clear when $A = 0$, we may assume that $A \neq 0$. Then $B \neq 0$ also, and $\alpha = \mu(B)$ is an infinite cardinal. Suppose that A has multiplicity α.

Since $A \lesssim B$, B has at least one nonzero submodule with multiplicity α. Let $\{B_i\}$ be a maximal independent family among those submodules of B which have multiplicity α. Then $B = E(\bigoplus B_i) \oplus C$ for some C, and the maximality of the family $\{B_i\}$ implies that C has no nonzero submodules with multiplicity α. As a consequence, none of the B_i are subisomorphic to C, whence $E(\bigoplus B_i) \not\lesssim C$. According to 6.22, $C \lesssim E(\bigoplus B_i)$, and so $B \lesssim 2E(\bigoplus B_i)$. Inasmuch as each B_i has multiplicity α, we infer that $E(\bigoplus B_i)$ must have multiplicity α. In particular, since $\alpha \geqq 2$ we see that $2E(\bigoplus B_i) \cong E(\bigoplus B_i)$, whence $B \lesssim E(\bigoplus B_i)$. Now 1.13 says that $B \cong E(\bigoplus B_i)$. But then B has multiplicity α, which is impossible.

Therefore A cannot have multiplicity α. Since α is infinite, we conclude that $\mu(A) \leqq \alpha = \mu(B)$. □

Proposition 6.24 Let $Z_r(R) = 0$, and assume that $S^{\circ}R$ is prime. If A is a nonzero nonsingular injective right R-module, then the following conditions are equivalent:

(a) A is directly finite.
(b) A does not have multiplicity 2.
(c) $\mu(A) = \aleph_0$.

Proof: (a) $\Rightarrow$ (b): If A has multiplicity 2, then $A \cong 2A$. Since $A \neq 0$, this contradicts 6.10.

(b) $\Rightarrow$ (c) is clear from the definitions.

(c) $\Rightarrow$ (a): If A is directly infinite, then there is an isomorphism g of A onto some proper submodule of itself. Now $A = gA \oplus C$ for some $C \neq 0$, and proceeding as in 6.10 we see that $C, gC, g^2C, \ldots$ are independent submodules of A. Then A has a submodule B which is an injective hull for $C \oplus gC \oplus g^2C \oplus \cdots$. Since $g^nC \cong C$ for all n, we see that $B \cong E(\aleph_0 C)$. Inasmuch as $(\aleph_0)^2 = \aleph_0$, it follows that B has multiplicity $\aleph_0$, and thus $\mu(B) > \aleph_0$ (since $B \neq 0$). According to 6.23, we obtain $\mu(A) > \aleph_0$, which contradicts (c).□

Proposition 6.25 Let $Z_r(R) = 0$, and assume that $S^\circ R$ is prime. If A,B are nonsingular injective right R-modules, then $\mu(A \oplus B) = \max\{\mu(A),\mu(B)\}$.

Proof: We may clearly assume that $A,B \neq 0$. According to 6.22, either $A \lesssim B$ or $B \lesssim A$; hence we may also assume that $A \lesssim B$. Note from 6.23 that $\mu(A) \leqq \mu(B) \leqq \mu(A \oplus B)$.

If B is directly finite, then A is directly finite by 6.8, and $A \oplus B$ is directly finite by 6.14. In this case 6.24 shows that

$$\mu(A \oplus B) = \aleph_0 = \max\{\mu(A),\mu(B)\}$$

If B is directly infinite, then by 6.24, B must have multiplicity 2, whence $A \oplus B \lesssim 2B \lesssim B$. Then 6.23 says that $\mu(A \oplus B) \leqq \mu(B)$, from which we conclude that $\mu(A \oplus B) = \max\{\mu(A),\mu(B)\}$.□

Corollary 6.26 Let $Z_r(R) = 0$, and assume that the ring $Q = S^\circ R$ is prime. If α is any cardinal, then the set $H(\alpha) = \{x \in Q \mid \mu(xQ) \leqq \alpha\}$ is a two-sided ideal of Q.

Proof: Since $\mu(0) = 0 \leqq \alpha$, we have $0 \in H(\alpha)$.

If $x \in H(\alpha)$ and $y \in Q$, then $xyQ \leqq xQ$ and $yxQ \lesssim xQ$. According to 6.23, $\mu(xyQ) \leqq \mu(xQ) \leqq \alpha$ and $\mu(yxQ) \leqq \mu(xQ) \leqq \alpha$, whence $xy, yx \in H(\alpha)$.

Given any $x,y \in H(\alpha)$, we have $(x - y)Q \lesssim xQ \oplus yQ$; hence $\mu((x - y)Q) \leqq \max\{\mu(xQ),\mu(yQ)\} \leqq \alpha$, by 6.23 and 6.25. Thus $x - y \in H(\alpha)$.□

Exercise 7 shows that 6.26 may fail if $S^\circ R$ is not prime. In the situation described in 6.26, we shall show that there are no other two-sided ideals of Q than the $H(\alpha)$. However, some preparatory results are needed first.

Lemma 6.27 Let $Z_r(R) = 0$, assume that $S^\circ R$ is prime, and let A,B be nonzero nonsingular injective right R-modules. If $A \not\lesssim nB$ for all positive integers n, then $A \cong E(\alpha B)$ for some infinite cardinal $\alpha \geqq \mu(B)$.

Proof: In particular, $A \not\lesssim B$; hence 6.22 says that $B \lesssim A$. Now let $\{A_i\}$ be a maximal independent family among those submodules of A which are isomorphic to B. Then $A = E(\oplus A_i) \oplus C$ for some C. Since $B \neq 0$, the maximality of the family $\{A_i\}$ implies that $B \not\lesssim C$. Then $C \lesssim B$ by 6.22, whence $A \lesssim E(\oplus A_i) \oplus B$. If α denotes the cardinality of the family $\{A_i\}$, we thus have $E(\alpha B) \lesssim A \lesssim E((\alpha + 1)B)$. By hypothesis, $\alpha + 1$ cannot be finite, whence α is infinite and so $\alpha + 1 = \alpha$. Then $A \lesssim E(\alpha B)$, and 1.13 shows that $A \cong E(\alpha B)$. If $\alpha < \mu(B)$, then since α is infinite we must have $A \cong E(\alpha B) \cong B$, which is impossible. Therefore $\alpha \geqq \mu(B)$.□

Theorem 6.28 Let $Z_r(R) = 0$, assume that $S^\circ R$ is prime, and let A,B be nonsingular injective right R-modules. Then $\mu(A) \leqq \mu(B)$ if and only if $A \lesssim nB$ for some positive integer n.

Proof: This is clear if either $A = 0$ or $B = 0$; hence we may assume that $A,B \neq 0$.

If $A \lesssim nB$ for some $n > 0$, then $\mu(A) \leqq \mu(nB) = \mu(B)$ by 6.23 and 6.25. If $A \not\lesssim nB$ for all $n > 0$, then by 6.27 there exists an infinite cardinal $\alpha \geqq \mu(B)$ such that $A \cong E(\alpha B)$. Inasmuch as α is infinite, we have $\alpha^2 = \alpha$, from which we conclude that A has multiplicity α. Therefore $\mu(A) > \alpha \geqq \mu(B)$.□

The simplest case of 6.28 says that if A and B are both directly finite and nonzero, then $A \lesssim nB$ and $B \lesssim kA$ for some $n,k > 0$. This can fail if $S^\circ R$ is not prime, as shown by Exercise 8.

Corollary 6.29 Let $Z_r(R) = 0$, assume that $S^\circ R$ is prime, and let A,B be directly infinite nonsingular injective right R-modules. Then $A \cong B$ if and only if $\mu(A) = \mu(B)$.

Proof: According to 6.24, A has multiplicity 2, whence $A \cong 2A$. As a result, we obtain $A \cong nA$ for all $n > 0$, and similarly $B \cong nB$ for all $n > 0$. If $\mu(A) = \mu(B)$, then by 6.28 $A \lesssim nB \cong B$ for some $n > 0$, and by symmetry $B \lesssim A$ as well. Then 1.13 shows that $A \cong B$.□

Theorem 6.30 Let $Z_r(R) = 0$, and assume that the ring $Q = S^\circ R$ is prime. Then every two-sided ideal of Q has the form $H(\alpha) = \{x \in Q \mid \mu(xQ) \leqq \alpha\}$ for suitable cardinals α.

Proof: Let H be any two-sided ideal of Q. If $H = 0$, then $H = H(0)$; hence we may assume that $H \neq 0$. Choose a nonzero element $x \in H$, and let α be the smallest cardinal such that $H \leqq H(\alpha)$.

If there exists an element $y \in H(\alpha) - H$, then $y \notin QxQ$; hence by 6.4 we have $yQ \not\lesssim n(xQ)$ for all $n > 0$. According to 6.27, there is an infinite cardinal $\beta \geqq \mu(xQ)$ such that $yQ \cong E(\beta(xQ))$. Inasmuch as β is infinite,

$\beta^2 = \beta$; hence yQ has multiplicity β. Since $y \in H(\alpha)$, we see that $\beta < \mu(yQ) \leqq \alpha$.

The minimality of α implies that $H \nleqq H(\beta)$; hence we may pick an element $z \in H - H(\beta)$. Since $\mu(zQ) > \beta \geqq \mu(xQ)$, 6.23 says that $zQ \nlesssim xQ$; hence we obtain $xQ \lesssim zQ$ from 6.22. Consequently,

$$yQ \cong E(\beta(xQ)) \lesssim E(\beta(zQ)) \cong zQ$$

According to 6.4, $y \in QzQ \leqq H$, which is a contradiction.

Therefore $H(\alpha) = H$.□

Theorem 6.31 If Q is a prime, regular, right self-injective ring, then the two-sided ideals in Q are well-ordered under inclusion.

Proof: Note from 2.11 that $Z_r(Q) = 0$ and $S^\circ Q = Q$. If X is any nonempty collection of two-sided ideals in Q, then by 6.30 there is a nonempty collection Y of cardinals such that $X = \{H(\alpha) \mid \alpha \in Y\}$. If β is the smallest cardinal in Y, then $H(\beta)$ is the smallest ideal in X.□

There is no requirement that the ideals $H(\alpha)$ in 6.30 be distinct. For example, since we have defined $\mu(xQ)$ to be either 0 or infinite, we always have $H(0) = H(1) = H(2) = \cdots = 0$. More generally, given any infinite cardinal β it is possible to have $H(\alpha) = 0$ for all $\alpha \leqq \beta$ (Exercise 9). Also, if γ is any cardinal such that $H(\gamma) = Q$, then obviously $H(\alpha) = Q$ for all $\alpha \geqq \gamma$. We can still ask whether the ideals $H(\alpha)$ between the first $H(\beta) \neq 0$ and the first $H(\gamma) = Q$ must be distinct. In order to prove that this is indeed the case, we must be able to construct principal right ideals xQ with specified values of $\mu(xQ)$, for which we use the following theorem.

Theorem 6.32 Let $Z_r(R) = 0$, assume that $S^\circ R$ is prime, and let A be any nonzero nonsingular injective right R-module. Let α be any positive cardinal, and set $B = E(\alpha A)$. If β denotes the successor of α, then $\mu(B) = \max\{\beta,\mu(A)\}$.

Proof: Since $A \neq 0$, $\mu(A) \geqq \aleph_0$. If α is finite, then $\beta < \mu(A)$, in which case we see from 6.25 that $\mu(B) = \mu(A) = \max\{\beta,\mu(A)\}$. Thus we may assume that α is infinite. If $\alpha < \mu(A)$, then since α is infinite we see that A has multiplicity α. In this case $B \cong A$, whence $\mu(B) = \mu(A) = \max\{\beta,\mu(A)\}$.

Now consider the case when $\alpha \geqq \mu(A)$. Since α is infinite, $\alpha^2 = \alpha$ and B has multiplicity α, whence $\mu(B) > \alpha$. Thus $\mu(B) \geqq \beta$.

Suppose that $\mu(B) > \beta$. Since β is infinite, B must have multiplicity β, and consequently $\beta A \lesssim \beta B \lesssim B$.

Choose a set X with cardinality β, and choose a module $C = \bigoplus_{i \in X} C_i$ such that each $C_i \cong A$. Since $C \cong \beta A \lesssim B$, there exists a monomorphism $f\colon C \to B$.

Now choose a set Y with cardinality α, and choose an independent family $\{B_j \mid j \in Y\}$ of submodules of B such that each $B_j \cong A$ and $E(\bigoplus_{j\in Y} B_j) = B$.

The family $\mathcal{F}$ of all nonempty finite subsets of Y is the union of the subfamilies $\mathcal{F}_n$, where $\mathcal{F}_n$ consists of all subsets of Y with cardinality n. Inasmuch as $\text{card}(Y) = \alpha$ is infinite, we infer that each $\mathcal{F}_n$ has cardinality α, and thus that $\text{card}(\mathcal{F}) = \alpha$.

For each $F \in \mathcal{F}$, set

$$X_F = \{i \in X \mid (fC_i) \cap (\bigoplus_{j\in F} B_j) \neq 0\}$$

Given any $i \in X$, we have $C_i \neq 0$ and hence $fC_i \neq 0$. Inasmuch as $\bigoplus_{j\in Y} B_j \leqq_e B$, it follows that $(fC_i) \cap (\bigoplus_{j\in Y} B_j) \neq 0$, from which we infer that $(fC_i) \cap (\bigoplus_{j\in F} B_j) \neq 0$ for some $F \in \mathcal{F}$, that is, $i \in X_F$. Therefore $X = \bigcup_{F\in\mathcal{F}} X_F$.

If $\text{card}(X_F) \leqq \alpha$ for all $F \in \mathcal{F}$, then since $\text{card}(\mathcal{F}) = \alpha$ we must have $\text{card}(X) \leqq \alpha^2 = \alpha$, which contradicts the assumption that $\text{card}(X) = \beta$. Thus $\text{card}(X_F) > \alpha$ for some $F \in \mathcal{F}$. Then $\beta \leqq \text{card}(X_F) \leqq \text{card}(X)$, whence $\text{card}(X_F) = \beta$. Set $D = \bigoplus_{j\in F} B_j$, and note from 6.25 that $\mu(D) = \mu(A) \leqq \alpha$.

Given any $i \in X_F$, we have $(fC_i) \cap D \neq 0$; hence $E_i = C_i \cap f^{-1}D$ is a nonzero $\mathcal{S}$-closed submodule of C_i such that $fE_i \leqq D$. Note from 6.2 that each E_i is injective. Choose an index $k \in X_F$ such that $\mu(E_k) \leqq \mu(E_i)$ for all $i \in X_F$. Setting $X_n = \{i \in X_F \mid E_k \lesssim nE_i\}$ for all $n = 1,2, \ldots,$ we see from 6.28 that $X_F = \cup X_n$. If $\text{card}(X_n) \leqq \alpha$ for all n, then we obtain $\text{card}(X_F) \leqq \alpha\aleph_0 = \alpha$, which contradicts the fact that $\text{card}(X_F) = \beta$. Thus $\text{card}(X_n) > \alpha$ for some $n > 0$. Then $\beta \leqq \text{card}(X_n) \leqq \text{card}(X_F)$, whence $\text{card}(X_n) = \beta$.

Now $E_k \lesssim nE_i$ for all $i \in X_n$, from which we see that $\beta E_k \lesssim n(\bigoplus_{i\in X_n} E_i)$. We have $fE_i \leqq D$ for all $i \in X_n$, and consequently $f(\bigoplus_{i\in X_n} E_i) \leqq D$. Since f is a monomorphism, we obtain $\bigoplus_{i\in X_n} E_i \lesssim D$, and thus $\beta E_k \lesssim nD$, whence $E(\beta E_k) \lesssim nD$. Observing that $E(\beta E_k)$ is a nonzero nonsingular injective module with multiplicity β, we see that $\mu(E(\beta E_k)) > \beta$. According to 6.25 and 6.23, it follows that $\mu(D) = \mu(nD) > \beta$, which contradicts our earlier observation that $\mu(D) \leqq \alpha$.

Therefore $\mu(B) \not> \beta$, and consequently $\mu(B) = \beta = \max\{\beta,\mu(A)\}$.□

Theorem 6.33 Let $Z_r(R) = 0$, and assume that the ring $Q = S^\circ R$ is prime. Then there exist infinite cardinals $\beta \leqq \gamma$ such that the rule

$$\alpha \mapsto H(\alpha) = \{x \in Q \mid \mu(xQ) \leqq \alpha\}$$

defines an isomorphism of the interval $[\beta,\gamma]$ onto the partially ordered set of nonzero two-sided ideals of Q.

Proof: According to 6.30, every two-sided ideal of Q has the form $H(\alpha)$ for suitable cardinals α. Let γ be the smallest cardinal such that $H(\gamma) = Q$. Since Q is prime, $Q \neq 0$; hence there also exists a smallest cardinal $\beta \leqq \gamma$ such that $H(\beta) \neq 0$. There is a nonzero element $x \in H(\beta)$, and $\beta \geqq \mu(xQ) \geqq \aleph_0$; hence β and γ are both infinite.

If $\beta \leqq \alpha \leqq \gamma$, then by 6.26 $H(\alpha)$ is a two-sided ideal of Q. Clearly $H(\beta) \leqq H(\alpha)$, whence $H(\alpha) \neq 0$. Thus the rule $\alpha \mapsto H(\alpha)$ defines a monotone map of $[\beta,\gamma]$ onto the set of nonzero two-sided ideals of Q. All that remains is to prove that if $\beta \leqq \alpha < \delta \leqq \gamma$, then $H(\alpha) < H(\delta)$.

Since $\alpha < \gamma$, $H(\alpha) < Q$, whence $\mu(Q) > \alpha$. Because α is infinite, it follows that $\alpha Q \lesssim Q$. We have a nonzero element $x \in H(\beta)$, and $\alpha(xQ) \lesssim \alpha Q \lesssim Q$; hence there must exist an element $y \in Q$ such that $yQ \cong E(\alpha(xQ))$. Inasmuch as $\alpha \geqq \beta \geqq \mu(xQ)$, 6.32 says that $\mu(yQ)$ is the successor of α. Then $\alpha < \mu(yQ) \leqq \delta$, and consequently $y \in H(\delta) - H(\alpha)$. Therefore $H(\alpha) < H(\delta)$.□

Exercises

1. Let F be a nonzero vector space over a field F. Prove that $[V:F] < \infty$ if and only if 1 is the only multiplicity of V.
2. Let F be a field, V an infinite-dimensional vector space over F, α any positive cardinal. Show that V has multiplicity α if and only if $\alpha \leqq [V:F]$.
3. If A is a module which has a multiplicity $\alpha \geqq 2$, prove that A also has multiplicity n for all positive integers n, and that A has multiplicity $\aleph_0$.
4. With R,B as in Exercise 3.A.4, show that R/B is directly finite but that $\mu(R/B) = \aleph_1$.
5. Let F be a field, $R = F \times F$. Find examples of R-modules $A \leqq B$ such that $\mu(A) > \mu(B)$.
6. Let F be a field, $R = F \times F$, α any uncountable non-limit cardinal. Find examples of R-modules A,B such that $\mu(A) = \mu(B) = \aleph_0$ while $\mu(A \oplus B) = \alpha$.
7. Given any infinite cardinal α, find an example of a right nonsingular ring R such that $H(\alpha)$ (as in 6.26) is not a two-sided ideal of $S^\circ R$.
8. Let $F_1,F_2,\ldots$ be fields, $R = \Pi F_n$. Find examples of nonzero nonsingular injective R-modules A,B such that A and B are directly finite but $A \nlesssim nB$ and $B \nlesssim nA$ for all $n > 0$.
9. Let F be a field, β any infinite cardinal, V a vector space over F of dimension β, $T = \mathrm{End}_F(V)$, M the (unique) maximal two-sided ideal of T, $R = T/M$. Show that $Z_r(R) = 0$ and that $Q = S^\circ R$ is prime. Show that the ideal $H(\alpha)$ of 6.26 is zero for all $\alpha \leqq \beta$.
10. Let A be a nonzero module. Show that the set of multiplicities of A has no maximal element if and only if $\mu(A)$ is an uncountable limit cardinal.
11. A cardinal β is *weakly inaccessible* if β is an uncountable limit cardinal and β cannot be expressed as the supremum of a set X of cardinals such that card(X) $< \beta$ and $\alpha < \beta$ for all $\alpha \in X$. (It is not known whether any weakly inaccessible cardinals exist.) If A is a nonzero module, prove that the set of multiplicities of A has no maximal element if and only if $\mu(A)$ is weakly inaccessible.

12. Let $Z_r(R) = 0$, assume that $S^\circ R$ is prime, and let A be a nonsingular injective right R-module. If there is a nonzero nonsingular injective right R-module B such that $\mu(B) < \mu(A)$, prove that $\mu(A)$ is not a limit cardinal.
13. Prove that every prime, regular, right self-injective ring is primitive.
14. Let R be regular and right self-injective. Prove that R is a direct product of prime rings if and only if every nonzero two-sided ideal in R contains a minimal two-sided ideal.
15. Let R be a prime, regular, right self-injective ring. If R is directly finite, prove that R is simple.
16. Prove 6.31 by using ideals $K(\alpha) = \{x \in Q \mid \mu(xQ) < \alpha\}$ in place of the ideals $H(\alpha)$ used in 6.30.
17. Let R be a simple, right self-injective ring. If R is directly infinite, prove that $R_R \cong xR$ for all nonzero $x \in R$.
18. Let R be regular and right self-injective. If there is an integer $n > 1$ such that $R_R \lesssim n(xR)$ for all nonzero $x \in R$ but $R_R \not\lesssim (n-1)(yR)$ for some nonzero $y \in R$, prove that R is simple artinian.
19. Let F be a finite field, β an infinite cardinal, R the free algebra over F on β letters. Using the notation of 6.26, prove that $H(\alpha) = 0$ for all $\alpha \leqq \beta$, while $H(\alpha) = S^\circ R$ for all $\alpha > \beta$.
20. Let $Z_r(R) = 0$, let A be a nonsingular injective right R-module, and set $T = \mathrm{End}_R(A)$. Prove that $\mu(fT) = \mu(fA)$ for all $f \in T$.
21. Given infinite non-limit cardinals $\beta \leqq \gamma$, find an example of a right nonsingular ring R such that β,γ are the cardinals given in 6.33.

Historical Notes

This section contains a series of brief notes, organized by chapter and section, relating the development of some of the ideas treated in the text. As with all such historical background, the choice of which ideas to discuss follows to a fair degree the author's preferences, modified by the availability of historical information. In certain cases—particularly with regard to terminology—a statement that x originated in a paper X should be interpreted as saying that the author was unable to locate any sources earlier than X in which x appeared. If any misattributions have occurred for this or other reasons, the author apologizes, and would welcome any corrections.

Section 1.A The concept of an essential submodule was introduced by Johnson [113, p. 891], without being named. The terminology is due to Eckmann-Schopf [44, 4.1].

Section 1.B Proposition 1.8 is due to Eckmann-Schopf [44, 4.2].

Baer [12, Theorem 4] proved the existence of minimal injective extensions [Theorem 1.10(b)]. Subsequently, Eckmann-Schopf [44, 4.1.4, 4.3] gave a much simpler proof by developing maximal essential extensions. The term "injective hull" seems to have been introduced by Rosenberg-Zelinsky [175, p. 373].

Theorem 1.13 was proved by Bumby [20, Theorem]. According to Faith [51, p. 21], this theorem was also proved independently by Osofsky (unpublished).

Section 1.C Two-sided versions of the terms "simple" and "semisimple" were introduced by E. Cartan [23, p. 1218], [24, ¶ 69], who defined a ring R

to be simple if 0 and R are the only two-sided ideals of R, and semisimple if R is a (finite) direct product of simple rings. These terms and definitions were later adapted to finitely generated modules, along with the alternate terms "irreducible" and "completely reducible." Krull [124, Definition 4, p. 64] subsequently expanded the definition of "completely reducible" to include modules which are not necessarily finitely generated.

The concept of the socle was introduced by Krull [124, Definition 5, p. 64], who called it the "largest completely reducible submodule." Remak [164, p. 4] defined the socle of a group to be the product of all minimal normal subgroups, and this name was adopted by Dieudonné [42, pp. 47, 51] in order to define the right and left socles of a ring.

The general Wedderburn-Artin Theorem given here (Theorem 1.18) evolved from a number of sources. The original idea is buried in a paper of Molien [143], who essentially proved that any finite-dimensional simple algebra over the complexes is isomorphic to a full matrix ring over the complexes. This result was later proved explicitly by E. Cartan [24, ¶71], who also showed that any finite-dimensional simple algebra over the reals is isomorphic to a full matrix ring over the reals, the complexes, or the quaternions [24, ¶85]. Wedderburn [208, Theorems 10, 17, 22] generalized this to arbitrary coefficient fields, by proving that any finite-dimensional simple algebra over a field F is isomorphic to a full matrix ring over a division algebra which is finite-dimensional over F, and also that any finite-dimensional semiprime algebra is a finite direct product of simple algebras. Subsequently, Artin [11] proved the analogous result for semiprime rings satisfying both ACC and DCC on right ideals, and then Hopkins [99, 6.4] showed that the DCC alone is sufficient. The arrangement of the theorem in terms of semisimple rings and modules [(a) $\Leftrightarrow$ (b) $\Leftrightarrow$ (d)] is due to Noether [146, pp. 663, 666, 667, 675]. A footnote attributed to the referee in Baer [12, p. 801] proved the equivalence (b) $\Leftrightarrow$ (f), and the equivalences (b) $\Leftrightarrow$ (e) $\Leftrightarrow$ (g) appeared in Cartan-Eilenberg [25, Theorem 4.2, p. 11].

Section 1.D Johnson [113, p. 894] defined the right and left singular ideals of a ring, and later [115, p. 537] defined the singular submodule of a module.

Section 2.A The localization functor S° is a special case of a localization procedure of Gabriel [65, Chapitre V, Section 2], which in turn is a special case of the construction of quotient categories due to Grothendieck [88, Section 1.11]. This particular case was developed by Gabriel [65, Chapitre V, Section 3].

The notion of an $\mathscr{S}$-closed submodule was introduced by Walker-Walker [205, Definition 4.10].

Goldman [75, Theorem 3.9] proved the left exactness of a class of localization functors which includes S°, and the right exactness of S° can be deduced from another result of Goldman [75, Theorem 4.5]. Theorem 2.6 in its present form was proved by the author [78, Theorem 1.14].

The regularity and self-injectivity of $S^\circ R$ (Proposition 2.11) were proved by Gabriel [65, Théorème 1, p. 418].

Section 2.B Quasi-injective modules were defined by Johnson-Wong [119, pp. 260, 261], who also proved Proposition 2.13 [119, Theorem 1.1].

Theorems 2.16 and 2.21 are the result of several improvements on an earlier result: If A is a nonsingular injective right R-module, then $\mathrm{End}_R(A)$ is a regular, right self-injective ring. This theorem is usually attributed to Johnson and Utumi, since the regularity follows from a result of Johnson [113, Theorem 2], while a proof due to Utumi [195, 4.6] can be used to prove the self-injectivity. The combined result appeared in Wong-Johnson [212, Theorem 5, Corollary to Theorem 4]. Theorem 2.16, for the case when A is injective, is due to Utumi [196, Lemma 8]. The full theorem was later proved by Faith-Utumi [53, Theorem 3.1], along with the result that if $J(Q) = 0$, then Q is right self-injective. The self-injectivity of $Q/J(Q)$ in general (Theorem 2.21) was proved independently by Osofsky [157, Theorem 12] and Renault [166, Corollaire 1 to Théorème 2.1], [167, Corollaire 3.5].

Section 2.C Rational extensions were introduced (but not named) by Utumi [195, 1.6]. The terminology is due to Findlay-Lambek [57, p. 81], who also proved the existence of maximal rational extensions (Theorem 2.26) [57, Theorem 2.6].

Utumi [195, 1.1] defined right quotient rings, and also defined and constructed maximal right quotient rings [195, pp. 3,4].

Section 2.D The term "abelian" is used here in the sense of Kaplansky [121, p. 10], while the term "totally non-abelian" is an invention for expository purposes.

Theorem 2.34 is a special case of a result of Utumi [197, Corollary to Theorem 4].

Theorem 2.35 is a modification of a theorem of Utumi [200, Theorem 1.4].

Theorems 2.37 and 2.38 are due to Utumi [197, Theorem 2], [200, Theorem 3.3].

Section 3.A Essentially finitely generated modules and essentially finitely related modules were introduced by Cateforis [26, Definitions 1.2 and 1.3], who also proved Theorem 3.5 [26, Lemma 1.5], [27, Proposition 1.5] and Theorem 3.6 [26, Theorems 1.9 and 2.1].

The equivalence (a) $\Leftrightarrow$ (b) in Theorem 3.9 was first proved in the commutative case by Akiba [1, Theorem 1]. The general case is due to Popescu-Spircu [161, Théorème 3.1], [162, Théorème 2.7].

Theorem 3.10 is due to Cateforis [27, Theorem 1.6].

The equivalence (b) ⇔ (c) in Theorem 3.12 was proved by Cateforis [27, Theorem 2.1]. The equivalence (a) ⇔ (c) was proved by the author [78, Corollary 2.6].

Section 3.B Finite-dimensional modules were introduced by Goldie [70, p. 590], [71, p. 202].

The equivalence (a) ⇔ (b) in Theorem 3.14 is due to Goldie [70, Lemma 1.1]. The equivalence (a) ⇔ (c) is due to Miyashita [142, Theorem 2.8].

Theorem 3.16 was proved by the author [78, Theorem 1.24].

The equivalence (c) ⇔ (d) in Theorem 3.17 is due to Gabriel [65, Lemme 6, p. 418]. The equivalence of (a), (b), (c), (e), (f) is due to Walker-Walker [205, Theorems 3.2, 3.5, and 4.19]. The equivalence of (d), (f), (g) was proved by Teply [191, Theorem 2.1].

Section 3.C Uniform modules were introduced by Goldie [70, p. 590], who also introduced the (Goldie) dimension of a module [71, p. 202].

Theorem 3.23 is due to Shock [181, Theorem 2.6].

Theorem 3.28 is due to Chase-Faith [35, Theorem 3.9].

Theorem 3.29 developed from several sources. Utumi [195, 5.1] proved that if R is primitive with nonzero socle, then $S^\circ R$ is a right full linear ring. Lambek [125, Theorem, p. 401] proved that if R is prime and $L^*(R_R)$ has at least one atom, then $S^\circ R$ is a right full linear ring. The implication (b) ⇒ (a) is due to Johnson [117, Theorem 3.1]. Subsequently, Hutchinson [103, Theorem 2] proved the equivalence (a) ⇔ (b), using the methods of Chase-Faith [35].

Section 3.D The equivalence (a) ⇔ (b) in Theorem 3.30 is due to Ore [150, Theorems 1,II].

Proposition 3.31 is contained in a proof of Mewborn-Winton [138, Theorem 2.5].

Theorem 3.34 is due to Goldie [71, Theorem 3.9].

The equivalence (a) ⇔ (b) in Theorem 3.35 is due to Goldie [71, Theorems 4.1, 4.4]. The equivalence (b) ⇔ (c) was proved by Sandomierski [176, Theorem 1.7].

A two-sided version of Corollary 3.36 was proved by Goldie [70, Theorem 13].

Section 4.A Formal triangular matrix rings have been used by a number of authors for constructing examples. Chase [34, p. 17] used them to construct an example of a ring which is left semihereditary but not right semihereditary. Small [184] used them to construct examples of right hereditary rings which are not left hereditary, and right noetherian rings which are not left noetherian. He also used this technique to find an example of a ring which is right hereditary but not left semihereditary [185, Section 1]. Herstein [97] used formal triangular matrix rings to construct an example of a right noetherian ring in which the intersection of the powers of the Jacobson radical is nonzero.

Theorem 4.7 is a special case of a theorem of Palmér-Roos [158, Théorème 1], [159, Theorem 5, p. 407].

Corollary 4.9 was proved by Eilenberg-Nagao-Nakayama [46, Proposition 12].

Section 4.B Notions of "essential irredundant subdirect sum" and "essential subdirect sum" very close to the definition of "essential product" were introduced by Chase-Faith [35, p. 258] and Hutchinson [101, p. 670]. Essential products were defined by the author [79, p. 493], who also proved Theorem 4.11 [79, Theorem 4].

Theorem 4.16 is a generalization of a result of the author [79, Corollary 5].

Theorem 4.17 is due to the author [79, Theorem 10].

Section 4.C The concept of the idealizer (of a principal left ideal) was introduced by Ore [151, Satz 12], under the name "Eigenring." Later, Fitting [59, p. 21] extended this to idealizers of arbitrary left ideals.

Proposition 4.19 is due to Robson [170, Proposition 2.1], [171, Corollary 1.5, Proposition 1.7].

Subidealizers were introduced by the author [85], who also proved Theorem 4.22 [80, Proposition 3], [85, Proposition 1.3].

Theorem 4.24 is a special case of a theorem of Robson [170, Theorem 2.8].

Theorem 4.26 is due to the author [85, Theorem 3.1].

Section 5.A The equivalence of (a), (c), (d) in Theorem 5.2 is attributed to Villamayor (unpublished) by Chase [33, Proposition 2.2].

Theorem 5.7 is contained in a theorem of Bass [13, Theorem P].

Eilenberg [45, p. 330] introduced the concept of a perfect category of modules, in which each module has a projective cover. Bass [13, p. 467] then defined a ring R to be right perfect if every right R-module has a projective cover, and proved that this definition is equivalent to the conditions of Theorem 5.7 [13, Theorem P].

Proposition 5.8 is due to Levitzki [134, Theorem 5].

Theorem 5.9 was observed by Bass [13, Example 1, p. 475].

Section 5.B Propositions 5.10 and 5.11 are due to Lenzing [131, Sätze 1,2].

Theorems 5.12 to 5.15 are due to Chase [33, Theorems 2.1, 4.1, 3.1, and 3.3].

Section 5.C Proposition 5.16 is contained in a result of Turnidge [194, Theorem 2.1].

Theorem 5.17 is due to the author [76, Theorem 7].

The equivalence (a) $\Leftrightarrow$ (b) in Theorem 5.18 is due to Cateforis [27, Theorem 2.3]. The equivalence (a) $\Leftrightarrow$ (c) was proved by the author [78, Theorem 2.5].

Proposition 5.19 and Corollary 5.20 are due to Sandomierski [177, Theorem 2.1, Corollary 2].

The implication (a) $\Rightarrow$ (c) in Theorem 5.21 was proved by Cateforis-Sandomierski [32, Proposition 3.2]. The remainder of the theorem is due to the author [78, Theorem 2.11].

The equivalence (a) $\Leftrightarrow$ (b) in Theorem 5.23, and the implication (a) $\Rightarrow$ (c), were proved by Cateforis-Sandomierski [32, Theorem 3.1, Proposition 3.2].

The statement of Theorem 5.28 was communicated to the author (without proof) by W. Stephenson. The proofs of Theorems 5.27 and 5.28 given here are due to the author [78, Theorems 2.14 and 2.15].

Section 6.A Theorem 6.3 is a slight generalization of a theorem of Renault [168, Théorème 1.1].

Theorem 6.5 is due to the author [82, Theorem 2.2].

Section 6.B Example 6.11 is due to Corner [39, Example 2].

Theorem 6.14, in the case when $A \oplus B = S^\circ R$, was proved by Renault [168, Proposition 5.1], using a method of Kaplansky [121, Theorem 56].

Theorem 6.16 was first proved in the case when $S^\circ R$ is prime by the author [81, Proposition 13]. The general case was proved by the author and Boyle [86, Theorem 3.8], using a method of Kaplansky [121, Theorem 36].

Proposition 6.18 was first proved in the case when A is injective by Warfield [206, Lemma 2]. The quasi-injective case is due to Fuchs [61, Theorem 3].

Lemma 6.19 was proved by Suzuki [188, Corollary 11]. The proof of Theorem 6.20 is due to Fuchs [62, Theorem 2], who proved that any quasi-injective module which satisfies the internal cancellation property of Lemma 6.19 must also satisfy the external cancellation property of Theorem 6.20. The observation that Lemma 6.19 combined with Fuchs' result yields Theorem 6.20 is due to Birkenmeier [16, Proposition 5].

Section 6.C Theorem 6.28 is a special case of a theorem of the author and Boyle [86, Theorem 14.9].

Theorems 6.30 and 6.31 are due to the author [81, Proposition 15, Theorem 8].

Theorems 6.32 and 6.33 are due to the author.

Bibliography

1. T. Akiba, "Remarks on generalized rings of quotients. III," *J. Math. Kyoto Univ.* **9:** 205–212 (1969).
2. J. S. Alin and E. P. Armendariz, "A class of rings having all singular simple modules injective," *Math. Scand.* **23:** 233–240 (1968).
3. J. S. Alin and S. E. Dickson, "Goldie's torsion theory and its derived functor," *Pacific J. Math.* **24:** 195–203 (1968).
4. S. A. Amitsur, "Rings of quotients and Morita contexts" *J. Algebra* **17:** 273–298 (1971).
5. F. W. Anderson and K. R. Fuller, *Rings and Categories of Modules*, New York: Springer-Verlag (1973).
6. E. P. Armendariz, "Quasi-injective modules and stable torsion classes," *Pacific J. Math.* **31:** 277–280 (1969).
7. ———, "On finite-dimensional torsion-free modules and rings," *Proc. American Math. Soc.* **24:** 566–571 (1970).
8. ———, "Q-divisible modules" *Canadian Math. Bull.* **14:** 491–494 (1971).
9. E. P. Armendariz and G. R. McDonald, "Maximal quotient rings and S-rings," *Canadian J. Math.* **24:** 835–850 (1972).
10. ———, "Q-divisibility and injectivity," *J. Algebra* **27:** 1–10 (1973).
11. E. Artin, "Zur Theorie der hyperkomplexen Zahlen," *Abh. Math. Sem. Univ. Hamburg* **5:** 251–260 (1927).
12. R. Baer, "Abelian groups that are direct summands of every containing abelian group," *Bull. American Math. Soc.* **46:** 800–806 (1940).
13. H. Bass, "Finitistic dimension and a homological generalization of semi-primary rings," *Trans. American Math. Soc.* **95:** 466–488 (1960).
14. R. L. Bernhardt, "Splitting hereditary torsion theories over semiperfect rings," *Proc. American Math. Soc.* **22:** 681–687 (1969).
15. ———, "On splitting in hereditary torsion theories," *Pacific J. Math.* **39:** 31–38 (1971).

16. G. F. Birkenmeier, "On the cancellation of quasi-injective modules," *Communications in Algebra* **4** (1976).
17. J.-E. Björk, "Rings satisfying a minimum condition on principal ideals," *J. reine angew. Math.* **236**: 112–119 (1969).
18. J. J. Bowe, "Neat homomorphisms," *Pacific J. Math.* **40**: 13–21 (1972).
19. B. Brainerd and J. Lambek, "On the ring of quotients of a Boolean ring," *Canadian Math. Bull.* **2**: 25–29 (1959).
20. R. T. Bumby, "Modules which are isomorphic to submodules of each other," *Arch. der Math.* **16**: 184–185 (1965).
21. A. Cailleau and G. Renault, "Sur l'envelope injective des anneaux de Baer," *C. R. Acad. Sci. Paris* **268**: 1381–1383 (1969).
22. ———, "Sur l'envelope injective des anneaux semi-premiers à idéal singulier nul," *J. Algebra* **15**: 133–141 (1970).
23. E. Cartan, "Sur les systèmes de nombres complexes," *C. R. Acad. Sci. Paris* **124**: 1217–1220 (1897).
24. ———, "Les groupes bilinéaires et les systèmes de nombres complexes," *Ann. Fac. Sci. Toulouse* **12**: 1–99 (1898).
25. H. Cartan and S. Eilenberg, *Homological Algebra*, Princeton: Princeton Univ. Press, (1956).
26. V. C. Cateforis, "Flat regular quotient rings," *Trans. American Math. Soc.* **138**: 241–249 (1969).
27. ———, "On regular self-injective rings," *Pacific J. Math.* **30**: 39–45 (1969).
28. ———, "Two-sided semisimple maximal quotient rings," *Trans. American Math. Soc.* **149**: 339–349 (1970).
29. ———, "Minimal injective cogenerators for the class of modules of singular submodule zero," *Pacific J. Math.* **40**: 527–539 (1972).
30. V. C. Cateforis and F. L. Sandomierski, "The singular submodule splits off," *J. Algebra* **10**: 149–165 (1968).
31. ———, "On commutative rings over which the singular submodule is a direct summand for every module," *Pacific J. Math.* **31**: 289–292 (1969).
32. ———, "On modules of singular submodule zero," *Canadian J. Math.* **23**: 345–354 (1971).
33. S. U. Chase, "Direct products of modules," *Trans. American Math. Soc.* **97**: 457–473 (1960).
34. ———, "A generalization of the ring of triangular matrices" *Nagoya Math. J.* **18**: 13–25 (1961).
35. S. U. Chase and C. Faith, "Quotient rings and direct products of full linear rings," *Math. Zeitschrift* **88**: 250–264 (1965).
36. A. W. Chatters, "PI-rings with zero singular ideal," *J. London Math. Soc.* **6**: 629–632 (1973).
37. T. J. Cheatham, "Finite-dimensional torsion-free rings," *Pacific J. Math.* **39**: 113–118 (1971).
38. R. R. Colby and E. A. Rutter, Jr., "The structure of certain artinian rings with zero singular ideal," *J. Algebra* **8**: 156–164 (1968).
39. A. L. S. Corner, "Three examples on hopficity in torsion-free abelian groups," *Acta Math. Acad. Sci. Hungary* **16**: 304–310 (1965).

40. J. H. Cozzens, "Homological properties of the ring of differential polynomials," *Bull. American Math. Soc.* **76:** 75–79 (1970).

41. S. E. Dickson, "A torsion theory for abelian categories," *Trans. American Math. Soc.* **121:** 223–235 (1966).

42. J. Dieudonné, "Sur le socle d'un anneau et les anneaux simples infinis," *Bull. Soc. Math. France* **70:** 46–75 (1942).

43. V. Dlab, "The concept of a torsion module," *American Math. Monthly* **75:** 973–976 (1968).

44. B. Eckmann and A. Schopf, "Über injektive Moduln," *Arch. der Math.* **4:** 75–78 (1953).

45. S. Eilenberg, "Homological dimension and syzygies," *Ann. of Math.* **64:** 328–336 (1956).

46. S. Eilenberg, H. Nagao, and T. Nakayama, "On the dimension of modules and algebras IV," *Nagoya Math. J.* **10:** 87–95 (1956).

47. D. Eisenbud and P. Griffith, "The structure of serial rings," *Pacific J. Math.* **36:** 109–121 (1971).

48. V. P. Elizarov, "Quotient rings," *Algebra and Logic* **8:** 219–243 (1969).

49. C. Faith, "Rings with minimum condition on principal ideals I," *Arch. der Math.* **10:** 327–330 (1959).

50. ———, "Rings with minimum condition on principal ideals II" *Arch. der Math.* **12:** 179–181 (1961).

51. ———, *Lectures on Injective Modules and Quotient Rings*, New York: Springer-Verlag, Springer Lecture Notes No. 49, (1967).

52. ———, *Algebra: Rings, Modules and Categories I* New York: Springer-Verlag, (1973).

53. C. Faith and Y. Utumi, "Quasi-injective modules and their endomorphism rings," *Arch. der Math.* **15:** 166–174 (1964).

54. ———, "Maximal quotient rings," *Proc. American Math. Soc.* **16:** 1084–1089 (1965).

55. D. J. Fieldhouse, "Note on the singular submodule," *Canadian Math. Bull.* **13:** 441–442 (1970).

56. K. L. Fields, "On the global dimension of residue rings," *Pacific J. Math.* **32:** 345–349 (1970).

57. G. D. Findlay and J. Lambek, "A generalized ring of quotients I," *Canadian Math. Bull.* **1:** 77–85 (1958).

58. ———, "A generalized ring of quotients II," *Canadian Math. Bull.* **1:** 155–167 (1958).

59. H. Fitting, "Primärkomponentenzerlegung in nichtkommutativen Ringen," *Math. Ann.* **111:** 19–41 (1935).

60. R. Fossum, P. Griffith, and I. Reiten, "Trivial extensions of abelian categories and applications to rings: an expository account," in *Ring Theory* (R. Gordon, ed.) New York: Academic Press, (1972) pp. 125–151.

61. L. Fuchs, "On quasi-injective modules," *Ann. Sc. Norm. Sup. Pisa* **23:** 541–546 (1969).

62. ———, "The cancellation property for modules," in *Lectures on Rings and Modules*, New York: Springer-Verlag, Springer Lecture Notes No. 246 (1972), pp. 191–212.

63. J. Fuelberth and M. L. Teply, "The singular submodule of a finitely generated module splits off," *Pacific J. Math.* **40:** 73–82 (1972).

64. K. R. Fuller, "On direct representations of quasi-injectives and quasi-projectives," *Arch. der Math.* **20:** 495–502 (1969).

65. P. Gabriel, "Des catégories abéliennes," *Bull. Soc. Math. France* **90:** 323–448 (1962).

66. P. Gabriel and U. Oberst, "Spektralkategorien und reguläre Ringe im Von-Neumannschen Sinn," *Math. Zeitschrift* **92:** 389–395 (1966).

67. E. R. Gentile, "On rings with one-sided field of quotients," *Proc. American Math. Soc.* **11:** 380–384 (1960).

68. ———, "Singular submodule and injective hull," *Indag. Math.* **24:** 426–433 (1962).

69. ———, "A uniqueness theorem on rings of matrices" *J. Algebra* **6:** 131–134 (1967).

70. A. W. Goldie, "The structure of prime rings under ascending chain conditions," *Proc. London Math. Soc.* **8:** 589–608 (1958).

71. ———, "Semiprime rings with maximum conditions," *Proc. London Math. Soc.* **10:** 201–220 (1960).

72. ———, "Rings with maximum condition," *Yale Univ. Lecture Notes* (1964).

73. ———, "Torsion-free modules and rings," *J. Algebra* **1:** 268–287 (1964).

74. ———, "Some aspects of ring theory," *Bull. London Math. Soc.* **1:** 129–154 (1969).

75. O. Goldman, "Rings and modules of quotients," *J. Algebra* **13:** 10–47 (1969).

76. K. R. Goodearl, "Embedding nonsingular modules in free modules," *J. Pure Applied Algebra* **1:** 275–279 (1971).

77. ———, "Distributing tensor product over direct product," *Pacific J. Math.* **43:** 107–110 (1972).

78. ———, "Singular torsion and the splitting properties," *Memoirs American Math. Soc.*, No. 124 (1972).

79. ———, "Essential products of nonsingular rings," *Pacific J. Math.* **45:** 493–505 (1973).

80. ———, "Idealizers and nonsingular rings," *Pacific J. Math.* **48:** 395–402 (1973).

81. ———, "Prime ideals in regular self-injective rings," *Canadian J. Math.* **25:** 829–839 (1973).

82. ———, "Prime ideals in regular self-injective rings. II," *J. Pure Applied Algebra* **3:** 357–373 (1973).

83. ———, "Triangular representations of splitting rings," *Trans. American Math. Soc.* **185:** 271–285 (1973).

84. ———, "Simple self-injective rings need not be artinian," *Communications in Algebra* **2:** 83–89 (1974).

85. ———, "Subrings of idealizer rings," *J. Algebra* **33:** 405–429 (1975).

86. K. R. Goodearl and A. K. Boyle, "Dimension theory for nonsingular injective modules," (to appear).

87. K. R. Goodearl and D. Handelman, "Simple self-injective rings," *Communications in Algebra* **3:** 797–834 (1975).

88. A. Grothendieck, "Sur quelques points d'algèbre homologique," *Tohoku Math. J.* **9:** 119–221 (1957).

89. R. N. Gupta, "Self-injective quotient rings and injective quotient modules," *Osaka J. Math.* **5:** 69–87 (1968).

90. D. Handelman, "When is the maximal ring of quotients projective?," *Proc. American Math. Soc.* **52:** 125–130 (1975).

91. ———, "Prime regular rings of quotients," *Communications in Algebra* **2:** 525–533 (1974).

92. D. Handelman and J. Lawrence, "Strongly prime rings," *Trans. American Math. Soc.* **211:** 209–223 (1975).

93. M. Harada, "Note on quasi-injective modules," *Osaka J. Math.* **2:** 351–356 (1965).

94. ———, "Hereditary semi-primary rings and tri-angular matrix rings," *Nagoya Math. J.* **27:** 463–484 (1966).

95. A. Hattori, "A foundation of torsion theory for modules over general rings," *Nagoya Math. J.* **17:** 147–158 (1960).

96. G. Helzer, "On divisibility and injectivity," *Canadian J. Math.* **18:** 901–919 (1966).

97. I. N. Herstein, "A counterexample in noetherian rings," *Proc. Nat. Acad. Sci. U.S.A.* **54:** 1036–1037 (1965).

98. ———, *Noncommutative Rings*, Math. Assoc. America, Carus Monograph No. 15 (1968).

99. C. Hopkins, "Rings with minimal conditions for left ideals" *Ann. of Math.* **40:** 712–730 (1939).

100. T. W. Hungerford, *Algebra*, New York: Holt, Rinehart, Winston (1974).

101. J. J. Hutchinson, "Intrinsic extensions of rings," *Pacific J. Math.* **30:** 669–677 (1969).

102. ———, "Essential subdirect sum decompositions of extension rings," *Arch. der Math.* **22:** 150–154 (1971).

103. ———, "Quotient full linear rings," *Proc. American Math. Soc.* **28:** 375–378 (1971).

104. ———, "Nonsingular Noetherian rings of exponent 2," *Math. Ann.* **209:** 267–276 (1974).

105. J. J. Hutchinson and J. Zelmanowitz, "Quotient rings of endomorphism rings of modules with zero singular submodule," *Proc. American Math. Soc.* **35:** 16–20 (1972).

106. ———, "Subdirect sum decompositions of endomorphism rings," *Pacific J. Math.* **47:** 129–134 (1973).

107. G. Ivanov, "Rings with zero singular ideal," *J. Algebra* **16:** 340–346 (1970).

108. ———, "Nonsingular rings with essential socle," *J. Austral. Math. Soc.* **17:** 358–375 (1974).

109. N. Jacobson, *The Theory of Rings*, American Math. Soc. Surveys No. 2 (1943).

110. ———, *Structure of Rings*, American Math. Soc. Colloquium Publ. Vol. 37 (1956).

111. ———, *Basic Algebra I*, San Francisco: W. H. Freeman (1974).

112. J. P. Jans, "Some aspects of torsion," *Pacific J. Math.* **15:** 1249–1259 (1965).

113. R. E. Johnson, "The extended centralizer of a ring over a module," *Proc. American Math. Soc.* **2:** 891–895 (1951).

114. ———, "Structure theory of faithful rings. I," *Trans. American Math. Soc.* **84:** 508–522 (1957).

115. ———, "Structure theory of faithful rings. II," *Trans. American Math. Soc.* **84:** 523–544 (1957).

116. ———, "Structure theory of faithful rings. III," *Proc. American Math. Soc.* **11:** 710–717 (1960).

117. ———, "Quotient rings of rings with zero singular ideal," *Pacific J. Math.* **11:** 1385–1392 (1961).

118. ———, "Rings with zero right and left singular ideals," *Trans. American Math. Soc.* **118:** 150–157 (1965).

119. R. E. Johnson and E. T. Wong, "Quasi-injective modules and irreducible rings," *J. London Math. Soc.* **36:** 260–268 (1961).

120. D. Jonah, "Rings with the minimum condition for principal right ideals have the maximum condition for principal left ideals," *Math. Zeitschrift* **113:** 106–112 (1970).

121. I. Kaplansky, *Rings of Operators*, New York: W. A. Benjamin (1968).

122. T. Kato, "Self-injective rings," *Tohoku Math. J.* **19:** 485–495 (1967).

123. L. A. Koifman, "Radikal modulya," *Sibirsk. Mat. Zh.* **7:** 1204–1207 (1966).

124. W. Krull, "Zur Theorie der allgemeinen Zahlringe," *Math. Ann.* **99:** 51–70 (1928).

125. J. Lambek, "On the structure of semi-prime rings and their rings of quotients," *Canadian J. Math.* **13:** 392–417 (1961).

126. ———, "On Utumi's ring of quotients," *Canadian J. Math.* **15:** 363–370 (1963).

127. ———, "On the ring of quotients of a noetherian ring," *Canadian Math. Bull.* **8:** 279–290 (1965).

128. ———, *Lectures on Rings and Modules*, Waltham, Mass.: Blaisdell (1966).

129. ———, *Torsion Theories, Additive Semantics, and Rings of Quotients*, New York: Springer-Verlag, Springer Lecture Notes No. 177 (1971).

130. J. Lawrence, "A primitive ring with nonzero singular ideal," *Proc. American Math. Soc.* **45:** 59–62 (1974).

131. H. Lenzing, "Endlich präsentierbar Moduln," *Arch. der Math.* **20:** 262–266 (1969).

132. ———, "Über kohärente Ringe," *Math. Zeitschrift* **114:** 201–212 (1970).

133. L. Lesieur and R. Croisot, "Sur les anneaux premiers noethériens à gauche," *Paris Ecole Norm. Sup. Ann. Sci.* **76:** 161–183 (1959).

134. J. Levitzki, "Solution of a problem of G. Koethe," *American J. Math.* **67:** 437–442 (1945).

135. L. Levy, "Torsion-free and divisible modules over non-integral domains," *Canadian J. Math.* **15:** 132–151 (1963).

136. ———, "Unique subdirect sums of prime rings," *Trans. American Math. Soc.* **106:** 64–76 (1963).

137. S. MacLane and G. Birkhoff, *Algebra*, New York: Macmillan (1967).

138. A. C. Mewborn and C. N. Winton, "Orders in self-injective semiperfect rings," *J. Algebra* **13:** 5–10 (1969).

139. G. Michler, "Klassische Quotientenringe von nicht notwendig endlich-dimensionalen halbprimen Ringen," *Math. Zeitschrift* **91:** 314–335 (1966).

140. R. Y. C. Ming, "A note on singular ideals," *Toho kuMath. J.* **21:** 337–342 (1969).

141. A. P. Mishina and L. A. Skornjakov, *Abelevi Gruppi i Moduli*, Moscow: "Nauka" (1969).

142. Y. Miyashita, "On quasi-injective modules, a generalization of the study of completely reducible modules," *J. Fac. Sci. Hokkaido Univ.* **18:** 158–187 (1965).

143. T. Molien, "Ueber Systeme höherer complexer Zahlen," *Math. Ann.* **41:** 83–156 (1893).

144. K. Morita, "Duality for modules and its applications to the theory of rings with minimum condition," *Sci. Rep. Tokyo Kyoiku Daigaku* **6:** 83–142 (1958).

145. ———, "Flat modules, injective modules and quotient rings," *Math. Zeitschrift* **120:** 25–40 (1971).

146. E. Noether, "Hyperkomplexe Grössen und Darstellungstheorie," *Math. Zeitschrift* **30:** 641–692 (1929).

147. K. C. O'Meara, "Primeness of right orders in full linear rings," *J. Algebra* **26:** 172–184 (1973).

148. ———, "Intrinsic extensions of prime rings," *Pacific J. Math.* **51:** 257–269 (1974).

149. ———, "Right orders in full linear rings," *Trans. American Math. Soc.* **203:** 299–318 (1975).

150. O. Ore, "Linear equations in non-commutative fields," *Ann. of Math.* **32:** 463–477 (1931).

151. ———, "Formale Theorie der linearen Differentialgleichungen (Zweiter Teil)," *J. reine angew. Math.* **168:** 233–252 (1932).

152. K. Oshiro, "On torsion-free modules over regular rings," *Math. J. Okayama Univ.* **16:** 107–114 (1973).

153. ———, "On torsion-free modules over regular rings. II," *Math. J. Okayama Univ.* **16:** 199–205 (1974).

154. B. L. Osofsky, "A semiprimitive ring with nonzero singular ideal," *Notices Amer. Math. Soc.* **10:** 357 (1963).

155. ———, "On ring properties of injective hulls," *Canadian Math. Bull.* **7:** 405–413 (1964).

156. ———, "A non-trivial ring with non-rational injective hull," *Canadian Math. Bull.* **10:** 275–282 (1967).

157. ———, "Endomorphism rings of quasi-injective modules," *Canadian J. Math.* **20:** 895–903 (1968).

158. I. Palmér and J.-E. Roos, "Formules explicites pour la dimension homologique des anneaux de matrices généralisées," *C. R. Acad. Sci. Paris* **273:** 1026–1029 (1971).

159. ———, "Explicit formulae for the global homological dimensions of trivial extensions of rings," *J. Algebra* **27:** 380–413 (1973).

160. N. Popescu, *Abelian Categories with Applications to Rings and Modules*, London: Academic Press (1973).

161. N. Popescu and T. Spircu, "Sur les épimorphismes plats d'anneaux," *C. R. Acad. Sci. Paris* **268:** 376–379 (1969).

162. ———, "Quelques observations sur les épimorphisms plats (à gauche) d'anneaux," *J. Algebra* **16:** 40–59 (1970).

163. C. Procesi and L. Small, "On a theorem of Goldie," *J. Algebra* **2:** 80–84 (1965).

164. R. Remak, "Über minimale invariante Untergruppen in der Theorie der endlichen Gruppen," *J. reine angew. Math.* **162:** 1–16 (1930).

165. G. Renault, "Etude des sous-modules compléments dans un module," *Bull. Soc. Math. France*, Mémoire No. 9 (1967).

166. ———, "Anneau associé à un module injectif," *C. R. Acad. Sci. Paris* **264:** 1163–1164 (1967).

167. ———, "Anneau associé à un module injectif," *Bull. Sci. Math.* **92:** 53–58 (1968).

168. ———, "Anneaux réguliers auto-injectifs à droite," *Bull. Soc. Math. France* **101:** 237–254 (1973).

169. ———, "Anneaux biréguliers auto-injectifs à droite," *J. Algebra* **36:** 77–84 (1975).

170. J. C. Robson, "Idealizer rings," in *Ring Theory* (R. Gordon, ed.) New York: Academic Press (1972), pp. 309–317.

171. ———, "Idealizers and hereditary noetherian prime rings," *J. Algebra* **22:** 45–81 (1972).

172. J.-E. Roos, "Locally distributive spectral categories and strongly regular rings" in *Reports of the Midwest Category Seminar*, New York: Springer-Verlag, Springer Lecture Notes No. 47 (1967), pp. 156–181.

173. ———, "Sur l'anneau maximal de fractions des AW*-algèbres et des anneaux de Baer," *C. R. Acad. Sci. Paris* **266:** 120–123 (1968).

174. ———, "Locally noetherian categories and generalized strictly linearly compact rings. Applications," in *Category Theory, Homology Theory and their Applications. II*, New York: Springer-Verlag, Springer Lecture Notes No. 92 (1969), pp. 197–277.

175. A. Rosenberg and D. Zelinsky, "Finiteness of the injective hull," *Math. Zeitschrift* **70:** 372–380 (1959).

176. F. L. Sandomierski, "Semisimple maximal quotient rings," *Trans. American Math. Soc.* **128:** 112–120 (1967).

177. ———, "Nonsingular rings," *Proc. American Math. Soc.* **19:** 225–230 (1968).

178. ———, "Some examples of right self-injective rings which are not left self-injective," *Proc. American Math. Soc.* **26:** 244–245 (1970).

179. D. W. Sharpe and P. Vamos, *Injective Modules*, Cambridge: Cambridge Univ. Press (1972).

180. J. C. Shepherdson, "Inverses and zero-divisors in matrix rings," *Proc. London Math. Soc.* **1:** 71–85 (1951).

181. R. C. Shock, "Polynomial rings over finite-dimensional rings," *Pacific J. Math.* **42:** 251–257 (1972).

182. ———, "The ring of endomorphisms of a finite-dimensional module," *Israel J. Math.* **11:** 309–314 (1972).

183. L. Silver, "Noncommutative localizations and applications," *J. Algebra* **7:** 44–76 (1967).

184. L. W. Small, "An example in noetherian rings," *Proc. Nat. Acad. Sci. U.S.A.* **54:** 1035–1036 (1965).

185. ———, "Hereditary rings," *Proc. Nat. Acad. Sci. U.S.A.* **55:** 25–27 (1966).

186. B. T. Stenström, *Rings and Modules of Quotients*, New York: Springer-Verlag, Springer Lecture Notes No. 237 (1971).

187. H. H. Storrer, "Rational extensions of modules," *Pacific J. Math.* **38:** 785–794 (1971).

188. Y. Suzuki, "On automorphisms of an injective module," *Proc. Japan Acad.* **44:** 120–124 (1968).

189. F. Szasz, "Über Ringe mit Minimalbedingung für Hauptrechtsideale. I," *Publ. Math. Debrecen* **7:** 54–64 (1960).

190. ———, "Über Ringe mit Minimalbedingung für Hauptrechtsideale. II," *Acta Math. Acad. Sci. Hungary* **12:** 417–439 (1961).

191. M. L. Teply, "Some aspects of Goldie's torsion theory," *Pacific J. Math.* **29:** 447–459 (1969).

192. ———, "Torsionfree injective modules," *Pacific J. Math.* **28:** 441–453 (1969).

193. K. Tewari, "Complexes over a complete algebra of quotients," *Canadian J. Math.* **19:** 40–57 (1967).

194. D. R. Turnidge, "Torsion theories and semihereditary rings," *Proc. American Math. Soc.* **24:** 137–143 (1970).

195. Y. Utumi, "On quotient rings," *Osaka J. Math.* **8:** 1–18 (1956).

196. ———, "On a theorem on modular lattices," *Proc. Japan Acad.* **35:** 16–21 (1959).

197. ———, "On continuous regular rings and semisimple self-injective rings," *Canadian J. Math.* **12:** 597–605 (1960).

198. ———, "On continuous regular rings," *Canadian Math. Bull.* **4:** 63–69 (1961).

199. ———, "On prime J-rings with uniform one-sided ideals," *American J. Math.* **85:** 583–596 (1963).

200. ———, "On rings of which any one-sided quotient rings are two-sided," *Proc. American Math. Soc.* **14:** 141–147 (1963).

201. ———, "A note on rings of which any one-sided quotient rings are two-sided," *Proc. Japan Acad.* **39:** 287–288 (1963).

202. ———, "On continuous rings and self-injective rings," *Trans. American Math. Soc.* **118:** 158–173 (1965).

203. ———, "On the continuity and self-injectivity of a complete regular ring," *Canadian J. Math.* **18:** 404–412 (1966).

204. ———, "Self-injective rings," *J. Algebra* **6:** 56–64 (1967).

205. C. L. Walker and E. A. Walker, "Quotient categories and rings of quotients," *Rocky Mountain J. Math.* **2:** 513–555 (1972).

206. R. B. Warfield, Jr., "Decompositions of injective modules" *Pacific J. Math.* **31:** 263–276 (1969).

207. ———, "Serial rings and finitely presented modules," *J. Algebra* **37:** 187–222 (1975).

208. J. H. M. Wedderburn, "On hypercomplex numbers," *Proc. London Math. Soc.* **6:** 77–117 (1908).

209. S. Wiegand, "Galois theory of essential extensions of modules," *Canadian J. Math.* **24:** 573–579 (1972).

210. C. N. Winton, "On the derived quotient module," *Trans. American Math. Soc.* **154:** 315–321 (1971).

211. E. T. Wong, "Rings with nonzero singular ideals," *J. Math. Kyoto Univ.* **10:** 419–432 (1970).

212. E. T. Wong and R. E. Johnson, "Self-injective rings," *Canadian Math. Bull.* **2:** 167–173 (1959).

213. A. Zaks, "Semiprimary rings of generalized triangular type," *J. Algebra* **9:** 54–78 (1968).

214. J. Zelmanowitz, "A shorter proof of Goldie's theorem," *Canadian Math. Bull.* **12:** 597–602 (1969).

215. ———, "Injective hulls of torsionfree modules," *Canadian J. Math.* **23:** 1094–1101 (1971).

Index

Arabic integers refer to page numbers, while Arabic decimals refer to theorems, propositions, etc. An expression of the form *m.X.n* refers to Exercise *n* in Section *X* of Chapter *m*.

Notation

A

B

C

D

E

J

L

M

N

O

P

T

U

V

W

Z